Hard Talk

Hard Talk

When Speech Is Difficult

Jonathan Cole

The MIT Press
Cambridge, Massachusetts
London, England

The MIT Press
Massachusetts Institute of Technology
77 Massachusetts Avenue, Cambridge, MA 02139
mitpress.mit.edu

The MIT Press would like to thank the anonymous peer reviewers who provided comments on drafts of this book. The generous work of academic experts is essential for establishing the authority and quality of our publications. We acknowledge with gratitude the contributions of these otherwise uncredited readers.

This book was set in ITC Stone Serif Std and ITC Stone Sans Std by New Best-set Typesetters Ltd. Printed and bound in the United States of America.

Library of Congress Cataloging-in-Publication Data

Names: Cole, Jonathan, 1951– author.
Title: Hard talk : when speech is difficult / Jonathan Cole.
Description: Cambridge, Massachusetts : The MIT Press, [2025] | Includes
   bibliographical references and index.
Identifiers: LCCN 2024022897 (print) | LCCN 2024022898 (ebook) |
   ISBN 9780262049566 (hardcover) | ISBN 9780262382045 (pdf) |
   ISBN 9780262382052 (epub)
Subjects: LCSH: Speech disorders—Patients. | Speech disorders—Patients—Means of
   communication. | Speech disorders—Patients—Psychology.
Classification: LCC RC423 .C586 2025 (print) | LCC RC423 (ebook) |
   DDC 362.19685/5—dc23/eng/20250109
LC record available at https://lccn.loc.gov/2024022897
LC ebook record available at https://lccn.loc.gov/2024022898

10 9 8 7 6 5 4 3 2 1

EU product safety and compliance information contact is: mitp-eu-gpsr@mit.edu

It is necessary to me not simply to be but to utter.

—George Eliot[1]

Contents

Introduction

We enter with a cry and, if fortunate, leave with a gentle exhalation or a sigh. In between, our lives are filled with chatter, from baby talk to old men's meanderings, one initiating and developing meaning, the other clinging onto it. Our lives and voices entwine, and it has been estimated that we speak round 500 million words in a lifetime.[1]

The refinement of chatter into increasingly sophisticated language has been cited as one of the main characteristics separating our species from others. The Oxford professor Max Muller thundered in 1861 that "Language is the Rubicon which divides man from beast."[2] In exploring what makes us special, the contemporary Oxford professor Richard Passingham concluded that humans were unique in their ability to speak, and in learning to do this naturally: "The essential human skill is that of speaking . . . teaching apes to speak is arduous, yet human infants pick up language at an astonishing rate and without directed programmes for teaching them."[3]

Though this somewhat underplays the parents' role, children's ability to absorb words is extraordinary; by age three most have a vocabulary of 1,000 words, by age six 13,000, and by eighteen 60,000. At the time of maximum acquisition, Robin Dunbar suggests, they acquire a new word every ninety minutes.[4]

There is a rich and fascinating literature on the scientific study of language and its structure, including semantics, word meanings, and syntax; how words are combined in phrases and sentences; studies of language as spoken, whether phonetics, the physical properties of speech sound production and perception, or pragmatics; how utterances are used in communicative acts; and the roles of context. One definition lists subgroups of investigation including sociolinguistics, dialectology, psycholinguistics, computational linguistics, comparative linguistics, and structural linguistics. These disciplines dissect the structures and commonalities of languages to peer at how deeply language is embedded into our cognitive functions, at the "how" and "what" of linguistic forms.

The evolution of language required massive structural rearrangement and repurposing of our upper respiratory tract, and before that the evolution of a more mobile face and tongue. Associated with these was reorganization and growth in the brain to control the action of speech, let alone the cognitive power underpinning language. It was hugely costly in evolutionary terms.

One of the first to consider language's purpose was the Russian neuropsychologist Lev Vygotsky, who approached this developmentally, from children's acquisition of language. "The primary function of speech is communication; social contact."[5] Vygotsky meant not simply that language's purpose is communication, but that this communication is itself primarily for and about social matters. "The earliest speech of the child is therefore essentially social."

Dunbar has reached similar conclusions. He has divided the evolution of language into three stages: contact and alarm calls, vocal chatter, and social shared meaning. His main purpose was to consider the evolutionary advantage of language. He thought that in the study of language insufficient attention had been paid as to why our species, alone, has evolved it. For Dunbar, the answer lies in how we use it. Humans are overwhelmingly—even addictively—social. Our conversations are as likely to be about who did what with whom than on the great matters of the day; most talk is social. Café talk is two thirds social; a similar fraction of book sales is fiction; while newspapers, especially the rags, would be all but empty without tittle-tattle. We are still coming to terms with the consequences of social apps enabled by smartphones and other online means of chatter.

His thesis is that language evolved under the social pressure to maintain relationships among members of one's group as that group size increased. He shows how nonhuman primates of various species have natural group sizes that increase, and as they do, the time spent in grooming each other increases too. Grooming's purpose, he suggests, is less to remove parasites than to reinforce social bonds. The problem is that as group size increases, the time to groom each other goes up so much that there is insufficient time for other essential matters, like foraging and sleep. The average group size of our ancestors is of course unknown, but Dunbar suggests it is related to the size of their brain neocortex, since larger groups need larger brains to navigate more complex social relationships.

The earliest hominids had groups of around 60, archaic humans 120, while our own natural group size is around 150–160 (based on evidence from a variety of sources, from the size of villages of Indonesian horticulturists to ideal church congregations and the numbers in military companies). To keep such a group together, he suggests, something other than physical grooming was necessary, and this something was language, evolution's solution to the need for more efficient social grooming. He suggests that for the same effort and energy we can reach three times more people through language than through physical grooming. Interestingly, with this development, grooming behavior became taboo except within a small group, such as perhaps one's family.

This is not to say, of course, that language remained purely social. It led to huge consequences for our cognitive and intellectual development, both through the spoken and written word, as well discussed by Passingham.

The various academic and clinical disciplines that have grown up around language were touched on above, but the primary focus of this work is talking rather than language, and talking revealed through the experiences of those for whom talk is hard. As the philosopher Ludwig Wittgenstein was forced to realize, between his first and second great books, language cannot be corralled by theory, but reveals itself through use and so through speech.[6] This field has become a part of linguistics and semiotics known as pragmatics, the study of how context and relationships, for instance, affect meaning as well as the linguistic contents of speech. The present work explores a tangential area: the difficulties in the uses of language by people with a variety of impairments in speech. Academic and even clinical studies have not always captured what it is like to live with speech impairments of various forms. First-person accounts are important for their own sake, validating individuals' experiences. As Oliver Sacks wrote,

> The study of disease, for the physician, demands the study of identity, the inner worlds that patients, under the spur of illness, create. But . . . these worlds cannot be comprehended wholly from the observation of behavior, from the outside. In addition to the objective approach of the scientist, we must employ an intersubjective approach, to see the world with the eyes of the patient himself.[7]

This approach should not remain solely the prerogative of physicians. According to the phenomenological philosopher Havi Carel,

> it is necessary to supplement a naturalistic account of disease, (physiological function) with a philosophical study of the experience of illness (how that disease is experienced) itself . . . to study illness without viewing it exclusively as a subject of scientific investigation. It is not enough to see illness as an entity studied with the tools of science. In order to fully understand illness it also has to be studied as a lived experience . . . to explore its existential, ethical and social dimensions.[8]

We may list the medical consequences of a problem with speech, but to fully understand its effects we need more detailed approaches. These allow us, often with force and passion, to realize the pervasive and sometimes unimagined ways in which vocal difference affects a person's life and social existence. Rarely is a voice problem only about the voice. What are the consequences for those whose voices may be muted or who find speech difficult, for whom it is hard to talk? Their experiences form the core of this work. This also allows us to reflect on and understand our own use of speech in new ways.

At one level, their losses allow insights into some of the mechanics of speech production—such as dysarthia (difficulty in speaking due to either the muscles used being weak or difficult to control)—which we normally take for granted. At the second level lie difficulties in the production and recognition of language, dysphasia, which occur with impairments of the language areas of the brain itself following stroke or tumor, or with the dissolution of neural function that characterizes dementia. These reach further into speech, language, and thought, and to how people view themselves. In order to understand these conditions from first-person perspectives, I have interviewed a series of people from disparate groups with various difficulties in speech.

In the first chapter, "No Zzzs: Acquired Vocal Cord Palsy," the experiences of two people with acquired vocal cord palsy are presented. The cords, through their vibration and extraordinary complex movements, add pitch, periodicity, and much more to our speech. The vocal cords also guard the lungs from inhalation of food and drink as well as being the narrowest point of the trachea. Perhaps surprisingly, this paralysis is not always recognized. It usually comes on in adulthood and can lead to an altered identity and early retirement. In those for whom paralysis is from birth, considered in chapter 2, "Words Unsaid: Congenital Vocal Cord Palsy," it may determine the sort of life a person leads. Chapter 3, "I Found Myself Boring: Vocal Cord Palsy," gathers other experiences of those with vocal cord palsy in relation to breathing, swallowing and conversation.

Chapters 4 and 5, "We Eat Eels Every Day: Spasmodic Dysphonia" and "Conversing with Myself: Spasmodic Dysphonia," consider the experiences of those who live with spasmodic dysphonia, a condition in which the neurological control of the vocal cords misfires, so that the cords may jam shut or open when talking (adductor and abductor spasm, respectively), though at other times, say, during breathing, their control is fine. One person described it as though her voice was being strangled by someone with a hand round her throat. Often doctors are unfamiliar with the diagnosis, leading to delays in treatment. Another person meditates on the relation between her intact internal dialogue and her fragmented communication with others through speech, an experience also related in several other conditions.

Above the voice box, or larynx, lie the nasopharynx, tongue, mouth, lips, and nose. The mobility of these organs allows the subtleties of sound required to make phonemes, those units of sound that distinguish one word from another. Chapter 6, "Stuff Happens: Cleft Lip and Palate," considers the mixed experiences of people with cleft lip and palate. These may lead not only to visible difference but also to problems with speech and hearing. Despite multiple operations, children with cleft may have to endure teasing and bullying. Others have suffered long-term abuse, assault, and even rape.

The rare congenital condition Möbius syndrome, examined in chapter 7, "Full On: Möbius," has two main characteristics: immobility of the muscles of facial expression with associated loss of eye closure or lip movement, so babies cannot suck and there is a continued need for eye care. The other main characteristic is that the eyes cannot be moved horizontally outward in divergence, leading to difficulties with distance vision. There can be associated problems with tongue and pharyngeal movements and with general clumsiness in limb and body movement. Unsurprisingly, children need more time and assistance to learn at school, but once this is in place there are no other learning problems.[9] Perhaps the key intervention, other than surgery, special teaching, and physiotherapy, is speech therapy. Certainly, Matthew Joffe, a man with Möbius now in his sixties and with many years of successful work in education and acting, thinks so:

I have had my share of surgeries, more than most, less than some. I have had eleven surgeries on my mouth, eyes, and hand; none of them rendered my "looks" normal in society's eyes.

The one intervention I had that truly empowered me was speech therapy. From the age of four in kindergarten to the end of ninth grade. Speech therapy was the one constant in my school life.

When I was three or four, I remembered the sound I made when I spoke. To this day, some sixty-plus years later, I can play this private tape in my head. I often describe it as a washing machine cycling in reverse, quite guttural and filled with water, the by-product of my excessive drooling.[10]

The main aim, he learned later, was to teach him to bring his voice up from the throat toward the front of the mouth. They also taught him to swallow before speaking so that the words were not ushered out "on a wave of water/saliva." He went to speech therapy once or twice weekly, one-to-one and then as part of group therapy with other children. So used to it was he that he thought it was part of being in school.

He was actually taught to employ a "tongue-tip placement system," using the tongue around the mouth to simulate the bilabial and fricative sounds that most of us make with our lips and mouths closed. An accomplished thespian who has sung, done stand-up comedy, and appeared in numerous documentaries around the world, Matthew notes that speech therapy transformed his life:

Group therapy allowed me to compete with other kids using my intellect and it boosted my confidence.

People within the Moebius community and elsewhere have congratulated me on how well I speak. It is one of greatest gifts anyone has ever given me, and I consider it one of my greatest achievements. It has opened doors and experiences in life that my less than Olympian body could ever afford me. The speech pathologists that were my mentors have my deepest gratitude. As a result of their work, I have been able to live a full life and embrace the world.[11]

In chapter 7, we meet Jo, who lives with Möbius too. The condition had extraordinary effects on Jo's sense of identity, with a reduced feeling of embodiment, so that as a child they did not feel their personhood was expressed within or through their own body. Jo was their mind and their thoughts, with their body, in contrast, just a collection of bits which let them down all the time. Even now in their forties, Jo has to concentrate before they express through words and gestures in a manner that others take for granted.

In his great work on the development of the human hand, published in 1833, the Scottish anatomist, artist, and neuroscientist Charles Bell realized that the anatomical differences between a human and many nonhuman primates' hands were small. Our improved dexterity was not primarily from low-level changes in the hand's hardware but due to increased capacities for manipulation and for planning of actions dependent on developments within our brain:

> Seeing the perfection of the hand, we can hardly be surprised that some philosophers should have entertained the opinion with Anaxagoras, that the superiority of man is owing to his hand. . . . Nevertheless [this] is not the cause of the superiority. Rather say with Galen, that man had hands given to him because he was the wisest creature. . . .
>
> With the possession of an instrument like the hand, there must be a great part of the organization which strictly belongs to it, concealed . . . A thousand intricate relations must be established throughout the body with it, nerves of motion and sensation; *there must be an original part of the composition of the brain before they can be put into activity.* . . . The hand would lie inactive, unless there were created a propensity to put it into action.[12]

Just as our unique use of the hands is dependent on our brain's evolution, so our vocal apparatus evolved in parallel with central developments for its control. It is often considered that the most developed and refined control of movement is reflected in the size of the representation of that body part in the sensorimotor cortex. Penfield's famous homunculus captures something of this with a vast representation of the hand and arm compared to that of other mammals. Yet just under half of both sensory and motor cortex is devoted not to the hand and fingers but to the face and vocal apparatus, with most related to the latter. In addition, there are many complex connections between the laryngeal motor cortex and surrounding areas in the somatosensory and parietal cortex for more complex sound production and its reception and understanding.[13] Large areas of our cerebral cortex are dominated by the need for speech production, reception, and processing.

Crucially, too, the laryngeal cortex has direct single-synapse connections with motor neurons in the brain stem that control the speech muscle and apparatus, in a similar way to the hand motor neurons within the hand motor cortex. This is a unique and new development in humans that reflects how important it is for the brain to control the muscles of speech precisely in the correct order and timing. When Passingham saw that these direct connections were found for speech muscles, he wrote, unusually for an empirical scientist and with surprising candor, that he felt "overjoyed."[14] This was powerful evidence that speech was different in man compared to all other nonhuman primates. The rest of this work considers some of the consequences of brain conditions that affect speech.

Parkinson's can be associated with tremor and movement, which can in turn become stiffened and difficult to initiate. But it is more than a movement disorder. Oliver Sacks wondered if the perception of time in people who have Parkinson's might also be altered. Speech is also affected by the condition, with talk slowing down and voices becoming more muted.

In chapter 8, "Just like My Elbows: Parkinson's," we meet Lizzie, a person with Parkinson's. This came on gradually, such that for some years she did not realize it was affecting her voice as much as her limb movement and gait. What surprised her too, over and above the problems with volume and voice production, was just how much more difficult the ebb and flow of social conversation was than her school conversation where the rules of turn taking and content were more formal. The chapter ends with the way she has found to mitigate some of the effects of Parkinson's on her loss of speech through an intensive form of speech therapy.

Cerebral palsy (CP), congenital and/or neonatal damage to the brain, may affect speech in variable, complex ways, even depriving some people of any means of expression. Christopher Nolan and Christy Brown both had such severe CP that they were unable to make useful, controlled movements for most of their lives, and their difficulties with controlling their vocal cords and tracts deprived them of speech. Despite that, they both wrote well-known and deeply affecting accounts of their lives.[15] Both sadly died in their forties from choking on food.

Cerebral palsy has a myriad of forms and of severities, and most people are not so severely affected as Nolan and Brown. Speech is possible for most people with CP, but even when it is, this does not mean they are at ease with its production. Chapters 9 to 11 are concerned with some of the problems that people with cerebral palsy can experience. Two concern a community of people with varying severities of CP and communication difficulties. Chapter 9, "In My Head, I Talk Slowly: Cerebral Palsy," relates the experiences of people with CP as they discuss how to be understood and how others respond to them. Chapter 10, "How We Live: Cerebral Palsy," focuses on two people with CP who have reduced or absent speech or other means of communication. The third chapter about CP focuses on a university lecturer in philosophy, whose cerebral palsy affects her head and neck, and thus her speech and swallowing apparatus. Chapter 11, "Acquiesce with Silence: Cerebral Palsy," considers the ways this limits not only her ability to express herself, but how her thoughts themselves are constrained by what she can express.

The following chapters concern one of the most frequent causes of speech difficulty: post-stroke aphasia. A major division is between those who have problems with speech but largely retain understanding, or "nonfluent aphasia," and those who are able to speak, but whose speech may have reduced intelligibility, or "fluent aphasia." People with this latter condition also have a reduced understanding of others' speech. To write a comprehensive account of the types of aphasia is beyond this work, and, in any case, there is a huge and excellent literature populated by giants such as Freud, Luria, Gardner, Damasio, and Pinker, as well as by many clinical neurologists, neuropsychologists, and speech therapists. Rather, this account focuses on how various types of aphasia affect a person's abilities to communicate with others and, as a result, how they see themselves.

In chapter 12, "Walks on the Wild Side: Nonfluent Aphasia," Steve, relatively soon after his stroke and with visual problems as well, and Bill, who several years later has made a partial recovery and feels he has become more relaxed and social, reflect on their new way of living. As we will learn, neither of their aphasias arrived alone.

Chapter 13, "Uncontrollable Dog: Nonfluent Aphasia," concerns Faye, a woman now in her forties who had a severe stroke leading to a nonfluent aphasia twenty years ago. Her recovery exceeded expectations of the medical staff. She describes what it was like to be without words and how they gradually returned. Even now, though, her speech and language remain affected.

In chapter 14, "Absence and Presence: Nonfluent Aphasia," we meet two men with nonfluent aphasias who have recovered less. They remain profoundly challenged in their communication. Their situation may be more easily observed than appreciated since they appear stranded, unable to express their thoughts. Their situations illuminate how identity and presence can depend on language and communication. And perhaps most of all we learn to appreciate that spouses and partners are affected by aphasia as well as their afflicted loved ones, albeit in different ways.

The fourth chapter on aphasia, chapter 15, "Eloquent Rhythms of Nothing: Fluent Aphasia," focuses on two men with fluent aphasia, in which speech or utterance returns with prosody and expression, but with little or no semantic content or meaning. In the first example, of Wernicke's aphasia, Paul has little awareness of his predicament, while the second had a rare form of aphasia leaving him without recognizable words at all, but with retained fluency, gesture, and prosody and with remarkable well-retained cognition. His stroke and its aftermath were explored by his wife, a notable author. Her extraordinary journey into her husband's utterances and internal world reveals much about the expressive, non-semantic components of speech and prosody while also revealing the limitations of utterance alone.

Chapter 16, "Nothing in Mind: Abulia," ends the narratives. It concerns a man who, after a very rare stroke, has been left with impoverished emotional experience. Though he has no deficit in speech production or perception, or in memory, this had severe consequences for one particular form of conversational speech.

The conclusion considers some additional cases of speech impairment, for instance, the case of those with dementia, in whom speech and language diminish, and those who stutter, or stammer. The chapter's main focus, however, is to draw out similarities in experience in those with the various speech problems—whether cleft, cerebral palsy, Parkinson's, or aphasia— examined in the preceding chapters. One of the surprising reports is that people think their voices sound normal. Why this might be the case is discussed, before moving to an account of various models of disability that, though theoretical, have bearing on the way in which difference is viewed.

We exist in dialogue. Again and again, impairments of talking, whether due to voice production or its central control, question the usual unbidden connections between linguistic thought, expression, and social existence. The aim of the present project is to explore these from the perspective of those for whom talk is hard. This may mean that making words is difficult, as in vocal cord paralysis; that machinery for this does not work properly, as in Parkinson's; that the link between word and meaning is loosened, or the link between thought and speech breaks down, as in some aphasias; or that the relations between language and thought and feelings itself become unfastened. We close by returning to the ways in which those with speech problems can be helped and what the rest of us should do to help.

A short postscript, "Inner Voices in Aphasia," considers recent work on two types of speech not designed to be heard by others, inner and private speech, in control subjects, and in those who sign their language and in those with aphasia. These reveal something of the ways used to understand those with communication difficulties.

To talk, to converse, is normally to be in relation to another or others. And though language can be technical and precise, it is often social and about such important matters as who did what, with whom, and when. One can analyze how language carries information and its complex structures, but chatting is as much about people getting on together as about words. Though immensely complex computer languages have been developed, computers don't yet make small talk or gossip. We *are* our appearance and our facial expressions; indeed, philosopher Maurice Merleau-Ponty wrote that we exist socially in the facial expressions of the other. But we are our speech as well, and, unlike these other means, speech is something we *do*; we talk of speech acts with an element of performance. *Hard Talk* explores this through the consequences of impairment in the voice. By looking at what happens when utterance becomes problematic, this work is about reaching out to others through voice and how this, in turn, is one of the most important ways in which we are defined.

1 No Zzzs: Acquired Vocal Cord Palsy

Put your hands together as though praying. Now turn the thumbs inward so the nails are facing each other and bend the tips of the thumbs apart, say 20 degrees or two centimeters. That is similar or slightly larger than the size and movement of the vocal cords.[1]

They sit in the larynx, between the throat and the trachea. Within the larynx are nine cartilages, three paired and three unpaired, which keep the tube rigid while also tilting to stretch the cords. Muscles within the larynx act either on these cartilages or more directly on the cords themselves. The cords open or shut and tense or relax in bewilderingly complex ways, and in extraordinary combinations of actions and speeds that depend on the neural control of the muscles as much as their anatomy.[2] The shape of the cartilages and the actions of the muscles involved in modulating vocal cord movement are so complex that it is difficult to show them in illustrations. Instead, there are a number of good videos that can be recommended.[3]

The nerve supply of these muscles is via two branches of a cranial nerve, the vagus, which snakes its way through the neck. The upper branch, the superior laryngeal nerve, supplies sensory fibers to the upper larynx and controls a muscle that tilts a cartilage to stretch the cords, used especially during higher-pitch sounds. The lower branch, the inferior laryngeal nerve, supplies the other muscles. It is known more widely as the recurrent laryngeal because though the right one takes a fairly direct route from the vagus to the larynx, the left one leaves the vagus far lower in the neck and descends to loop right around the aorta before ascending toward its target. Such a long route makes it vulnerable to inflammation and surgery to that area.

Voicing

The cords open during breathing, shut when swallowing, and move during speech. The cords also shut during actions requiring increased rigidity in the upper body and chest: defecation, weightlifting, and even, it is said, during orgasm. Breathing, of course, primarily allows transfer of gases: oxygen in and carbon dioxide out. During speech we tend to override these needs, with longer exhalations to fit our words and sentences. But equally, when we need gas transfer—say, while running—that need predominates. Then speaking in long, measured sentences becomes difficult.[4] We breathe in by actively expanding the chest wall and deepening the diaphragm; breathing out depends on the elastic recoil in the lungs. This is carefully controlled during speech to allow air to pass through the vocal tract according to our needs for vocalization.

The cords can open and shut at rapid rates, under neural control, to control vocal pitch. Estimates vary, but the rates are 90–500 times per second in men and 150–1,000 times per second in women, with speech using rates of around 120 in men and 200 in women. They can also open when higher air pressures force them open. Once open and the pressure is reduced, they spring back. This turns a smooth breath into interrupted packets of air, heard as a buzz: "zzz" (with the tongue pressed against the upper palate), which can be heard and indeed felt if you place fingers on the larynx. The cords also alter in tension and position and can move and vibrate at high rates to alter the sound reaching the later parts of the vocal tract. Sounds are usually the result of complex movements of lungs, vocal cords, tongue, palate, and lips. Those sounds produced primarily by vocal cord movement are called voicing. Though they require upper tract movement at their beginning, they are sustained by vibration of the cords, as in the sounds of "b," "d," "g," "v," and, of course, "zzz."[5]

After passing through the cords, air then reaches the rest of the vocal tract, the throat or pharynx behind the tongue, the mouth with its mobile tongue and palate, and the nose, and then it is released through the lips. Each chamber is a different shape and size and can move in different ways, which all affect sound. It is remarkable that our voices are so individual. Some of this may reflect the different architecture of our vocal tract. Just as we look different, we all sound different.

The sounds we hear to distinguish words are called phonemes. Vowels tend to be air passed through the tract without obstruction, with the different sounds being produced by movements of the tongue, jaw, and lips. Consonants are produced by absolute obstruction of the outflow—plosives or nasals—or by narrowing of it—fricatives— or by both, the affricates. Plosives are produced by stopping the airflow using the lips, teeth, or palate, followed by a sudden release of air. Some sounds like "t," "k," and "p" are done with the front part of the tract, with relatively smooth air flow, and are termed "voiceless," while "d," "g," and "b" require a more sudden release of the air, dependent on vocal cord movement, and so are "voiced." A nasal consonant, as the name implies, involves air leaving through the nasal cavity, usually with the lips, as in "m," or tongue, as in "n," blocking air through the mouth. An affricate is produced by stopping air leaving the tract and then releasing it suddenly, as in the "ch" sound.

Different languages have developed different phonemes apparently randomly. An added complexity is that as we move from one word to another at speed, each phoneme may differ, being slightly dependent on those immediately before and after, as the vocal tract moves between them. What we hear is anyway a deduction or inference, since while we hear distinct words their sound is produced more continuously.

Paralysis

Vocal cord paralysis is rare and can occur due to immobility of the cartilages on which the cords sit or due to impairment of the nerves supplying the vocal muscles. It can follow intubation, when the cartilages or nerves are damaged, surgery to the neck, or can occur for no clear reason, which is often put down to viral inflammation. Though babies are vulnerable during intubation, most cases occur in adults. The overall incidence is 0.42 percent, with most developing it in their fifth or sixth decades (77.2 percent). The onset of symptoms was gradual in 60 percent of cases, with the left vocal cord affected twice as commonly, its nerve supply making it more vulnerable. The cause was unclear and presumed to be inflammation or an unknown virus in just under 40 percent, with cancer next most common (30 percent).[6] Vocal cord paralysis after surgery for thyroid problems occurs in around 3–5 percent of people postoperatively, with 0.3–1 percent being permanent, and can also be seen after surgery on the chest, heart, or aorta, again due to the long course of the left recurrent laryngeal nerve.[7]

When the superior laryngeal nerve is affected, the main symptom is a loss of voice for higher-pitched sounds, though there are also swallowing problems and abnormal sensation in the upper larynx. Recurrent laryngeal nerve damage leads to more vocal problems, with a rough or breathy voice, reduction in range and speed, as well as tiredness on talking and swallowing problems. Much depends on where the cord lies when paralyzed. If one cord is paralyzed near the midline, then the other can compensate and close the cords when necessary. Then voice and swallowing are less affected. If the cord lies open, closure of the cords is impossible and the voice is affected more and the risk of choking increased. Such medical descriptions, however, do not fully describe the experiences of those with vocal cord palsy or consider its consequences in the long term.

Tess: "My New Normal"

About seven years ago, when she was age fifty-three, Tess went to her general practitioner because she was short of breath. Presuming asthma, he prescribed a steroid inhaler. It seemed to help, though in retrospect this was probably because she would sit down and rest for a while after taking it. Her voice changed too, which was put down to Hashimoto's thyroiditis, an immune condition of the gland in the neck.

Her shortness of breath soon increased such that she could barely shower or make it up stairs. She went to a lung doctor, who admitted her immediately and scoped her with a laryngoscope. Both vocal cords were paralyzed. She was referred to one ear, nose, and throat (ENT) specialist, and then another more experienced in vocal cord palsy. He suggested an operation to remove part of one vocal cord, or a cordotomy. This would improve her airway, but might reduce her speech (since the cords might not close together to vibrate properly), and there was an increased risk of aspiration during swallowing. This was all too new for Tess; she wanted to investigate all possibilities before irrevocable surgery.

She trawled the internet and found that Larry Page, then CEO of Google, had paralyzed cords after a viral infection and also had Hashimoto's thyroiditis. She found her way to his surgeon, who had a hall of fame of pictures on his office walls. He thought Tess was remarkably similar to Page. It turned out that Page had donated money to find an alternative to cordotomy, and the surgeon had a treatment on trial. Tess was not convinced

and went back to her original surgeon. He gave her three choices: a cordotomy, a permanent tracheotomy (opening in the front of the trachea
with insertion of a small tube to allow her to breathe through it), or—even
less appealing—death. "Choose one. You cannot keep walking around like
this." Liking his bluntness, Tess chose a limited vocal cordotomy.

Since then, she lives what might be considered "a pretty much normal
life"—or at least her new normal life—but remains out of breath during
almost any activity:

> This has had the most negative impact on my life. I cannot do the things I would
> like. I just cannot keep up with other people. My husband does most of the clean-
> ing. Even just dusting gets me out of breath.

A little while after the operation, she retired:

> Most everyone in the office was younger than myself and many of my coworkers
> were guys. I was the manager and had to keep these young guys in line. I started
> using the hand up in a STOP way so that I could finish my thought.

They were actually very sympathetic and understanding. A bunch of
jokers, they never saw her as less than before because of the voice issue.
"They actually made fun of me and joked with me quite a bit—all in a good
way. Having that job was a blessing." Indeed, her boss had not wanted her
to have the tracheotomy because it would "change who she was." Her biggest fear was a respiratory infection, so if anyone at the office was sick, she
worked from home. Even a runny nose made it much harder to breathe
because of her restricted airway.

She avoids large parties, even weddings, "because you just know someone will be ill." She travels less and no longer flies. "I like the comfort of
home with my humidifier, my air purifier, all my vitamins and my essential
oils. It's just made my world a bit smaller, I guess." In the family, if anyone
is sick they reschedule get-togethers for when everyone is feeling better.

She rarely aspirates food, though sometimes drinking liquids or swallowing large pills is awkward. Oddly, it sometimes happens if she thinks too
much about swallowing. And despite her best intentions, she panics more.
Small spaces and elevators are worst. "I'm not sure why but the air feels
harder to breathe in these spaces. I have to get out of there *now*." She also
finds it difficult to have house guests:

> They just feel too close. I do not sleep well with other people staying over. We
> have a summer home and frequently have guests both there and at our main

residence. I will wake up feeling I cannot breathe and start to panic. I need to escape but with nowhere to go. I enjoy friends and family, so I choose to deal with the panic instead of not having the company.

Her breathing problems improved following the cordotomy but remain a fact of life to be tolerated and endured. Vocal problems too: "I don't have much voice volume. If you are in the other room from me, you will not hear me. I cannot yell." She keeps a whistle with her in case she needs to get anyone's attention quickly. She is aware that when in a quiet place her breathing is louder than others' and that she has to clear her throat more frequently.

Without volume, it is hard in loud places like bars and restaurants: "I will use my phone to type out my order and show the waitperson." Her voice immediately "drops to the ground when out of my mouth." She finds it difficult to perform an uplift at the end of a sentence to ask a question, her voice a breathy monotone. Some days "b"s are hard, other times "m"s have a "b" sound to them. Every day is different.

I cannot do "z"s, of course. They come out as an "s" sound. ZOO is SOO. This is how I explain that my vocal cords are paralyzed, especially to children. If you feel the side of your throat when you say "z" you can feel it vibrate. There is no movement on mine. It is hard to explain this condition to people because they think if your vocal cords are paralyzed then how can you talk. This can help them understand.

With the change of voice, she feels, has come a change in her identity. She used to talk to anyone and everyone; now she is quieter, less outspoken, and always trying to think how to say something with fewer words before someone interrupts. She even enjoys quiet more. Alone in the house she no longer has music or the TV on just for noise. But it has not altered how she views herself: "I'm still the same person, for me if not for others."

The loss of natural conversation is huge! By the time I take my breath to get my words out in a conversation, people have already started their sentence. Then I can appear rude if I go ahead and start my sentence, as if interrupting. More times than not I just exhale and listen. I must say, this is especially aggravating when you want to interject a joke or a funny one-liner.

Interacting with someone in a casual way becomes strange. I don't speak to casual strangers as much I as used to. For instance, waiting in a room with people, there is a kid you want to interact with. . . . I do not do that much anymore. Everyday conversations are not natural; even in the line at the grocery store, it is a whole thing. People think you are sick. I cannot tell you how many times I say, "No, I'm not sick . . . my vocal cords are paralyzed."

With friends and family, it is easier.

> I hold up my hand in a STOP sign . . . and they know to let me finish my sentence. That's all I want; to just finish my sentence. It is sometimes hard to get out everything I am thinking. Just because of how many breaths it takes just to express an entire thought. I can only get out about seven or eight words before I must take another breath. Sometimes I just don't want to try.

Other times she has to forcefully interrupt to ask people to let her finish her sentence. It made her realize how some people do not actually listen to the other person in a conversation. They are clearly not really listening if only half her sentence made sense.

More recently, she has begun to use a small amplifier with friends. Initially shy about using it in public, people have been supportive:

> As soon as you explain why you are using it—and when they hear me talk—people get it and think it's a great idea. Before I would just not talk very much in loud places. I would just sit back and listen.

Conversations are especially hard on the phone and so she prefers to text, even with her ninety-seven-year-old mother, who is hard of hearing anyway. Indeed, texting each morning and evening has brought them closer.

The vocal cords and tract do not just articulate language; with limited vocal cord function Tess finds it difficult to convey emotional range:

> I can't laugh out loud. My laugh is gone. That tone—to laugh—I cannot get to it. It's not in my range. I laugh but nothing comes out. My shoulders heave when I am cracking up, but no sound comes out. I miss that. A lot. I try not to cry anymore. Because if I cry things swell up inside and make it hard to breathe. So, I shed a tear or three and move on.
>
> I've been married to my husband for twenty-seven years. I have wondered, If he met me now would he still be interested in me? I was so talkative before. I still am but it just isn't the same. Our sense of humor is part of what attracted us; without the laughing out loud would it have been an issue?

Now that it is physically exhausting to carry on long conversations, she has become a better listener. She also misses singing along with her favorite songs. "I know it may sound silly, but many of us [with vocal cord paralysis] miss this simple little thing." She says:

> I mourned for the person I was and have moved on. I don't like to feel sorry for myself. When people give me the sad, poor-you face, I explain that my vocal cords are paralyzed. Whenever I get that pity look . . . I quote Regina Brett, "If we all threw our problems in a pile and saw everyone else's, we'd grab ours back." It is not the end of the world. It's just my new normal.

Elaine: Welcome to My World

Three years ago, Elaine, then aged fifty, could stand the nodules on her thyroid no more since they were beginning to affect her swallowing. The only option was removal of the gland. In the recovery room after the operation, she remembers her noisy breathing and that she nearly choked on some water. But things settled and she left hospital, though on thick liquids only and with a different voice. At her follow-up visit, the surgeon pronounced the surgery a great success.[8] She ordered a routine swallow study, which Elaine passed, and so she was sent on her way.

About two weeks later she returned, tired and short of breath. Previously, she had been jogging, playing tennis, and practicing yoga; now she could not climb stairs. The surgeon scoped her and found bilateral vocal cord paralysis, a known complication of the operation.

> Living with this was such a change. Before I would come home from work, put on running gear and just run, and not think about anything; 5K four times a week was perfect.

An athlete all her life, she kept trying to exercise, thinking it would help recovery. Like Lance Armstrong cycling through his chemotherapy for testicular cancer, she reasoned that if she could run she was fit, and that things would improve.[9] Even when her exercise tolerance was a miserable one to two minutes and she was gasping like she was dying, she was more embarrassed than anything. She kept going, aiming for five seconds per day more, small precious increments for a runner.

Her paralyzed vocal cords were resting in what is called the paramedian position, slightly apart. This meant that she could breathe, speak, and swallow, but none were good.

She is an occupational therapist, helping those with various physical conditions in activities of daily living. Talking is a big part of the job, and her voice was "terrible." She was always short of breath talking and needed two to three breaths to finish a sentence. She would write instructions on a board. This was not only frustrating—"a lot of tears"—but it was very tiring. An amplifier was not a success; its poor voice quality prevented understanding.

After nine months there had been no improvement. Elaine had learned of laryngeal nerve reinnervation and suggested it to her doctor. She was referred to a surgeon who was just starting a program. She was his second patient. Apart from the usual pre-surgery workup, she also had her vocal muscles recorded to check that the problem was due to damage to their nerve supply, and was scoped to make sure the joints in the larynx still moved (if not, she would not have been a candidate for the procedure). A month later she went back for the reinnervation. They created a new pathway around the damaged recurrent laryngeal nerve for the motor signals to the muscle that opens the vocal cords. The surgery took six hours. Elaine woke to a temporary tracheotomy and lots of tubes: oxygen, pulse oximeter, wound drain, IV, urinary catheter, and nasogastric tube. Her right ear was numb because they removed a nerve for the donor nerve. She had a repeat swallow study after six days to check all was well, and then was allowed home.

The quality and loudness in her voice began to improve about five months later. She had better closure of the cords, so air did not escape so much. She expected it would take one to two years for the nerve to regenerate and her voice to improve. Typically, she tried jogging about two months post-op with a friend. They would jog a couple of minutes and then walk about three or four minutes while her breathing recovered:

> It is frustrating. . . . I unrealistically hoped to return to prior level of function. But I am also grateful that I can do a lot of things now. Running is definitely the hardest—too bad because it is my favorite. Riding a stationary bike is easier (though when I say easy, I do not mean easy. . . . I still get short of breath but less). It is difficult to grade, but I use a ten-point scale. . . . 1 being extremely easy, 5 being somewhat comfortable but becoming challenging, and 10 being maximum effort, I was close to 8–9/10 with any activity before and now the bike is about 5/10, jogging about 6–7/10.

She now runs three to four miles three days a week, sometimes nonstop. Compared to where she was after the thyroidectomy, she is okay with that. She misses singing too, especially at her church but also in the car and in the shower.

To see if there was more she could do, she consulted a speech and language therapist. An enigma is that we do not actually know how we speak—it just happens—so what to do to improve speech remains beyond explanation too. We make sounds without really knowing what we are doing; we don't feel how our cords or the vocal muscles are working. Most feedback comes from what we hear. Elaine's therapist, Jim, related how sometimes people do try to direct their effort to the vocal folds, but this can then hamper rather than help, either by affecting the usual coupling between respiratory and vocal muscles or by attention to the laryngeal muscles rather to the whole vocal tract, leading to tension in the neck.[10]

Jim concentrated on three components:

1. Forward focus or resonance. Better projection of the voice through the anterior vocal tract, mouth, nose, and lips. Just as in projecting the voice on stage, this aimed at involving the mouth and lips more especially to sound consonants and exaggerate.

2. Improve flow of air over the folds. Elaine had intermittent and poorly controlled flow, leading sometimes to the voice becoming broken and thready with loss of clear phonation, like a stalled kite. By thinking about diaphragmatic control and smooth expiration, flow over the folds was improved and hence her voice improved too.

3. Improving breath economy. This is related to point 2 but involved relaxing into longer inspiration and longer speaking between breaths. Then she had fewer breaths, which were quieter and less frantic.

Jim was very aware that it is hard work to think about diaphragmatic control, breaths in, and anterior projection, while also listening to the other person and deciding what to say and whether it is easy to say. At the end of a day Elaine would be tired out and her throat hurt. Following therapy with less maladaptive focus on the larynx, she found a better, louder, deeper, more resonant voice. It even allowed sufficient voice to control the dog in the garden, where previously he had ignored her.

Like Tess, she finds some words, letters, and sounds more difficult. With a new client, in order to run a quick screen of their cognition, Elaine asks them to remember three words: sock, blue, and bed. But these particular words are hard for her to pronounce: "I need to put much effort into words ending in 'ck' and 'd,' and those starting with 'm.' If I try to be more expressive and get excited, then I lose volume and have increased breathiness." She uses more gestures and exaggerates her lip movement to help patients read her lips.

Getting back to work was a huge effort, and despite being in a caring profession some nurses made fun of how she sounded. Many times she would cry, unable to talk as she wanted and exhausted from talking and breathing. "What a thing we just take for granted." She thinks people, unaware of the effort she puts in, just presume that she has become less social.

I think people see a false calm . . . if I don't say anything or exert too much, I appear normal. People that know me, some of my co-workers and best friends, say I am strong. . . . But believe me I do not always feel so.

At work she used to be fun, though if there was a disagreement she felt strongly about, she never went down without a fight. Now she might express her view a couple of times and then drop the issue. One manager just kept talking over her, and she had to give up. She also finds a mismatch between thought and talk, her mind faster than her speech. "Then, my words come out jumbled, muffled . . . unclear. I try again . . . slower and focused. Add trying to control breathing and it equals a lot of thought and effort."

At home after work, Elaine is now so tired she talks little; she needs down time to conserve her energies and her voice. At work we tend to talk more slowly and deliberately, with better projection than when we are gossiping with friends or family. She rarely goes out for dinner with friends anymore because of the background noise. "Nobody went out for dinner during COVID-19. That is like my world since the palsy."

Some years ago, I interviewed a young man who had just lost his sight in his twenties.[11] Still coming to terms with this, he thought the lucky ones were those who had never seen. He reasoned that grieving over something lost was worse than never having experienced it. Both Tess and Elaine were adults when their condition began and could reflect on their time before and after their loss. What might the experience be of someone who had known no other existence, who had been born with vocal cord problems? I talked next with Meg.

2   Words Unsaid: Congenital Vocal Cord Palsy

Very rarely vocal cord paralysis is present from birth, say, due to a neurological impairment usually of the brain stem. It can follow intubation, particularly in premature babies, due to damage to either the laryngeal cartilage or the nerves. Most babies recover. If not, then treatment is similar to adults, with fillers, which are injected into the vocal cords to increase their size and allow them to meet better in the midline, cordotomy, and if necessary, tracheotomy. Since language acquisition in children begins in the second year, and long-term memories begin to be laid down some time after that, such paralysis is experienced as though congenital.

Meg: "As Long as I Stay Quiet, My Voice Is Normal"

Now nearly eighty, Meg was given radiation treatments for what was thought to be an enlarged thymus as a baby. When she started kindergarten in September 1950, the school nurse noticed a lump in her neck. It was a cancer of the thyroid gland, which was removed in two operations. After the second, her problems with paralysis of her vocal cords began. She remembers little before this, so knows no other existence.

A tracheotomy was followed with a year out of school. At home there was little need to talk, though she remembers putting her finger over the tracheotomy hole if she wanted to say anything. Her parents took turns sleeping in her room in case she could not breathe at night. After fourteen to fifteen months, she returned to school with the tracheotomy. She has a photo a year later at a Brownie Scout meeting. The other girls are smiling, joking, and acting out; Meg is just sitting there with fairly flat affect. "I was there but not a part of the group. I was a mostly silent observer." In 1954 she had arytenoid surgery, during which one of her arytenoid cartilages was reduced in size (these are pyramid-shaped cartilages at the back of the larynx). The surgery improved her vocal cord position and helped her breathing, after which her tracheotomy could be removed. Further treatment for the remaining cancer followed with radioactive iodine in 1957.

Being out of class for over a year and then returning with a tracheotomy tube meant she missed out on group activities. Instead, there were doctors' visits, hospital stays, family time, and home schooling. She recalls just one visit from a school friend.

Catching up was difficult. Prior to returning to school, Meg's mother had visited the class to prepare them. When Meg arrived, no one talked about it. This was fine, like it was no big deal, but in reality for Meg it was. She started fourth grade after arytenoid surgery with a new voice, and had to try to fit back in. "I hated reading aloud in class. I did a lot of just watching and little speaking. Why bother?" Her mother did her best to present it all as normal: "La-de-da—we're off for another trip to the hospital or the doctor," with the best intentions of downplaying the disruptions, but it meant that Meg did not chat about it and had no opportunity to process, let alone express, what she was feeling. Only later did she realize how much had passed her by. She heard her voice around age eleven or twelve on a tape recorder. "I hated it." She dreaded her French language lab in high school because she had to record and listen to her voice. She found speaking French especially difficult perhaps because of phrasing/breathing issues when trying to pronounce a foreign language.

Meg spent her elementary school years in the relative safety of a small class in a small school. In contrast, the junior high school was large. By then she had learned to fit in, her goal to pass as normal.[1] She even met a few friends outside school. But what she really enjoyed was making music. She began playing flute in bands, marching at football games and in parades, and playing with the orchestra for school theatrical performances. At last, she had a group to hang out with:

> I wanted to play flute and march in the band. Of course, my mother said I could never do that. But I insisted! I played and marched all through high school. Band people were my people. Sustaining long notes on flute was always a challenge but I made it work. It helped with breathing and voice control, allowing better phrasing when I spoke. The breathing techniques I learned for flute help me sneak in little breaths between words when speaking.

She was a good student and did well academically, though she remained a silent observer. "All through school, I longed to be part of the in-group: partying, dating, and being popular." As before, she kept these feelings to herself. "A high school classmate said I was incredibly quiet but spoke up when I had something I wanted to say. That was an accurate observation that fits even today."

She went on to a local state college, and had to take one required speech class. She picked the absolute worst for her: oral interpretation, which involved reading poems and essays out loud. At the end of the class, the instructor told her she would not qualify as a teacher because of her speech. Though she had never wanted to be a teacher, she was upset for failing at something even though she was unaware of being so evaluated. She would have been better speaking her own words, tailored to her limitations, rather than someone else's.

Meg dated during college and married in 1965, with three children following. Her social sphere was based around family. She preferred one-on-one friendships to larger group activities. It was difficult to read to her children, or even yell at them, since she would run out of breath, so she gave up both and they managed. She went back to college and graduated, majoring in psychology. When the children were in their teens, she and her husband separated. She had become more of a home person, while he pursued his love of boating and social activities at the local yacht club. Though nothing was said, the vocal palsy played its part.

Once single, Meg needed a better job. Even with her degree she had lots of interviews but no offers. Finally, someone mentioned she sounded like a breathless, ditzy blond. Only then did she realize her voice was holding her back. In 1986, she became an analyst for the state of California with help from an affirmative action officer. She worked in vocational rehabilitation, in personnel and as an auditor. Such analytical work suited her very well. The job also boosted her confidence, with her voice less of a problem. Most of her interactions could be with small groups in quiet rooms.

Meg has lived her whole life with breathing problems and a quiet, breathy voice. Strangers have asked so often if she has a chest cold that she has to hide her annoyance. She breathes shallowly, and can be short of breath walking, not helped by now living at altitude. Her big fear has always been respiratory infections. New doctors do not always understand that her vocal cord paralysis means she cannot cough properly, making her more prone to these. Indeed, it took her years to understand that herself. The transition from being a dependent child to an independent adult, albeit one who needs some extra care, was difficult. As a child, her parents navigated her treatment and care; as an adult she had to develop new support teams herself.

She can only speak two to four words per breath, so she speaks in clips, with quick breaths between clusters of words. "What time—do you want—to eat?" or "I need to go—to the bank.—Do you want—to come with me?" With more words the air runs out, leaving words unsaid.

With a complicated thought, she may practice in her head before breaking it into clusters. Speaking quickly, she may sneak in an extra word, but if nervous she slows down. Changes in the last decades have been so gradual that she is unsure of them:

> I have always had to think about this limitation. It affects my ability to express complicated thoughts, especially if stressed. Sometimes when emotional, my voice just gives out and I need to start over. I cannot count on it and can even have trouble pronouncing my name. I must try ridiculously hard for people to hear "Meg" and not "Mat."

Meg does not remember much about her speech patterns as a child. The only childhood speech therapy she remembers is shouting vowels A-E-I-O-U during her one and only session.

She knows her voice has affected her identity most of her life. "I'm quiet and reserved and seen to be that way by others. My affect is a bit flat most of the time. Always the silent observer."[2] Even now, her voice limits her ability to express herself. She has to think about what she says beforehand:

> All the time: I have a thought pattern: What do I want to say? What are the most important components? How much noise is in the room? How far do I need to project? Will I be heard? Can I say enough quickly enough to be fully understood? Is it worth the effort? Practice it in my head, do I need to watch how I say it for the best effect? Do I expect disagreement? Go for it or abort?

Rehearsing and composing so much in her mind, she sometimes does not remember whether she said something or just thought it. "A lot of what I think never gets expressed."

Natural conversation has never been easy or fluent. As a result, she has been thought aloof when she was just trying her best. She has always avoided phone calls, speaking on answering machines, and attending large group functions. "As long as I stay quiet, my voice is normal."

One area where she can escape the voice problem is making music. "Playing music allowed me to express myself in ways speaking just never did." Abandoning her flute after high school, she later joined a New Horizons Band for Seniors while living in Northern California. Breathing was difficult, but she played well enough to keep up. On moving to Utah, all of the New Horizons groups were string orchestras, so she learned cello. A string instrument was a better fit. She loves expressing herself musically on it, freed from breathing and voice limitations, being like everyone else:

> This group has given me a role among a group of friends which highlights an issue with my speech. Talking comes much more easily when enthusiastic about what I want to say and comfortable with my audience. I express myself faster, with more words per breath and with less rehearsal in my head. I don't have to think about words; they just come out.

She has become leader of the orchestra, something she would never have dreamt of years ago. She speaks up to get her message out, and the group has learned to listen to her. "I just do it. My friendships within the group have allowed me to become more comfortable speaking out to them."

It is similar with family. In friendly and familiar settings, thoughts and words are in harmony and in real time. If not quite like everyone else, it is still liberating compared with much of her life. In contrast, debating remains difficult:

> When trying to prove a point or settle a disagreement, my head goes in circles trying to pick out the best few points to express. I get stuck. I am intimidated easily and just back down or retreat.

Speech is an intensely social act and depends so much on context and on the confidence to engage with others, which we often take for granted, forgetting the adolescent torments of speaking in public. For many with impairments, talking—thereby encouraging attention toward oneself and what one has to say—requires not simply confidence but courage as well. Meg's observations show how great the differences in fluency, articulacy, and engagement are for her in various social conditions, with support or with potential conflict.

By contrast, writing and email have given Meg an outlet for her thoughts freed from voice and difficulty:

> Written communication has become one of my strengths. Especially on a computer, my thoughts just race out through the keyboard. Sometimes in a hurry, I

just put down phrases. They are not always full sentences. But when needed, I can weave things together to get my point across.

In the past she has sought treatments hoping for a better voice and breathing all over the country including both the East and West Coasts. But these have just made her realize that her original surgeon gave her the best compromise between breathing and speaking. "What they did way back in 1954 has never needed to be redone. I guess it will never get any better than that." But the limitations are never far away and she is constantly monitoring herself. "I am more aware of my breathing limitations and my speech patterns. I find myself making less and less effort to express myself verbally." Think of how we chat, gossip, joke, and josh, with no effort or thought involved in how we speak. Meg knows how much her paralysis has cost. Despite this, she is now in her seventies with three grown-up children and seven grandchildren. She has reached an accommodation with her situation. "My paralysis played a huge role in my social development and life, but it is okay."

3   I Found Myself Boring: Vocal Cord Palsy

In the course of writing this account of mute voices, I approached many support groups. Some of the stories from those groups concerning vocal cord problems lent themselves to extended narratives, but others, like the stories from Martha, David, Hannah, Tess, and Marjolein, were better as short examples of the problems posed by the condition.

Breathing

We are not usually aware of or even think about breathing; if we had to, then sleep for one thing would be difficult.[1] Athletes, of course, get out of breath as they push themselves and show the typical heaving chest and wide-open mouth as they force air in and out. In asthma the tubes of the lungs constrict, airflow becomes audibly wheezy, and the patient can become panicked and desperate. An acute attack is a terrifying medical emergency for all concerned. Less dramatic is the shortness of breath seen in chronic lung damage, say, emphysema from smoking, when so much lung tissue has been lost that there is insufficient tissue for transfer of gases in and out, and permanent breathlessness results. These people may become aware of their breathing, but for most of us, most of the time, it just happens.

In vocal cord paralysis, due usually to damage to the cartilage supporting the vocal cords or to the nerves, the cords are often stuck in a position between fully open or fully shut. This reduces the aperture of the larynx and so reduces airflow. Sometimes this is insufficient for normal breathing at rest, as Tess described, while for others such as Elaine it becomes a problem for anything that increases the need for air, such as exercise. Another person with vocal cord paralysis, Martha, explains:

I am fine sitting, fine walking on the flat, but pant hard walking up even moderate hills. I can manage a single flight of stairs but have to pause on a second. Yoga gets me breathless. I tried to start running again and felt like I was going to pass out.

I self-identify as a runner. But I haven't run for almost two years and may not run again. I like walking and walk 5–6 K every day, so I'm pretty much okay with this in terms of identity but it's still weird. Anyway, runners are always having injuries that stop them running, so plenty of so-called runners are not running at any given point. My seventeen-year-old son is now running regularly and my fourteen-year-old daughter has just started. I've loved watching her go from very resistant to running to starting a short jog and building up to an hour run each morning before school getting up and going it completely on her own. I would so, so love to run with either one of them. But I can't.

Running might be considered inessential, even luxurious, but as Martha says, it is part of the identity of an increasing number of people for both fitness and social reasons. It also speaks to an important limitation with vocal cord paralysis: that any increase in activity, whether climbing stairs or taking a shower or walking the dog, can lead to breathlessness and discomfort. It imposes a uniformity to activity that can neither be circumvented nor escaped.

A further constant concern is the propensity to infection, which can produce excess mucus, reduce the airway size, and tip people into problems. As we have seen, the fear of chest infections was one of the main reasons that led Tess to retire, a fear that drives many with cord paralysis further from social interaction, whether at work or with friends and even family.

It is not always the big things, either. Meg describes a nagging tickle and cough (improved by occasional Botox injections, which she avoids since they soften her voice). Coughing occasionally is accepted, but when frequent it can be annoying and upsetting. Many people with vocal cord palsy also make small, involuntary, gasping noises that are embarrassing and draw unwanted attention.

Swallowing

We think as little about swallowing, normally, as we do about breathing. It occurs through a complex series of reflex activities as food is chewed into a squidgy small mass, or bolus, mixed with saliva and then passed to the back of the mouth. Sensory receptors in the posterior mouth and pharynx lead to reflex closure of the mouth, nasopharynx, and vocal cords, and relaxation of the upper esophagus to allow food to pass down into the stomach. Avoidance of aspiration—food going down the wrong way into the lungs—is through closure of the vocal cords and is controlled via the recurrent laryngeal nerve. This nerve is also the pathway for the complex sensory information from the pharynx and larynx that initiates reflex activity. Damage to the cords or cartilage compromises reflex cord closure, while damage to the nerve can affect these and the sensation in this area initiating the normal swallow too.

Those with vocal cord paralysis are constantly on the alert; swallowing is no longer a given. Some prefer to avoid foods with both solid and liquid content. For many, diet is determined by how easy a food is to swallow. David, who has a vocal cord palsy of unknown cause, wrote:

> I am careful with food; small bites to prevent choking, no dry food, like biscuits, muesli, dried fruit, or crackers. I always make sure that food is wet, that is, sauce, gravy, and other wet things on food. I drink lots of water during meals.

Another volunteered that she can choke on crumbly things like cereal or on lumpy soup. Meg finds tilting her head back when drinking can bring it on and can send her into coughing fits.

During a swallow and vocal cord closure, breathing is interrupted, so for some with breathing problems, eating adds to their difficulties. Breathing and eating may be taxing, but trying to eat, breathe, and talk are even more of a problem. What for most of us is one of the more pleasant, shared necessities of life—food and drink—may become a daily trial.

Conversation

Hannah's left vocal cord paralysis followed removal of her left thyroid lobe. This was never mentioned as a possible side effect, and she had never heard of it:

> Being told your vocal cord is paralyzed is so brutal and final. A part of you not working, never working again, permanent. It was dropped like a brick on me. Years later, my head may have accepted it, but my heart still finds it difficult.

She tried everything, determined to recover, with speech therapy ("nine nuns knitting on the moon") to strengthen her vocal cord muscles, various surgical options, Botox, reinnervation, and a filler injected into the cord all suggested. By then she was pregnant and other priorities took over. Still years later, her voice

> drives me bloody mad. It is a big chunk out of my well-being. I am not how I was. I cannot do anything to get it back. I cannot fix it. I cannot do anything.

She grieves for her lost voice and finds living with a low-volume, thin voice difficult. It is such an effort she feels as though she is shouting all the time. Talking takes energy and is tiring, with aching behind the jaw, down into the throat, even around the collarbone. The paralysis has affected two big parts of her life: being a mum and having a career. She could not talk quietly or whisper to her newborn son. Trying to avoid clearing her throat or coughing when holding a sleeping baby was never easy. At work she went from being a successful manager in healthcare to struggling to hold down a basic part-time administrative job. She used to be keen to chair meetings; now she never volunteers and avoids casual chat. She knows it gives her new colleagues the wrong impression and is isolating, but cannot help it.

Our voices are not just about words; speech prosody, intonation, emphasis, and loudness add richness, interpretation, and personality. As Marjolein wrote, "I can express my feelings, though it is difficult to express enthusiasm or encouragement with my flat voice."

Martha, who has a left vocal cord paralysis due to recurrent laryngeal nerve damage, describes the losses her voice problems have brought:

> I can't really read to my youngest now; reading aloud takes a lot of breath. He and I used to read to each other at breakfast every day. Now I can't read while I eat. I choked when I tried. This is a genuine loss. I haven't read to him at bedtime for eighteen months and at nine he's almost outgrown it now.

Voice is for communication, for exchanges between people:

> There has been a loss of natural conversation. In one-to-one talking it's okay, but groups are hard. I have to make myself do group social events and I've not been bothering. I can't participate properly; I feel people are humoring me if they talk to me (like chatting to someone's granny). It's hard to talk to a stranger at a group event long enough to find that common ground that makes you both want to continue.
>
> It's also hard to find new friends when you feel it's difficult for them to get to know you. Most events are in a group or at a coffee shop or restaurant where I can't be heard. It would be very odd to ask a new potential friend straight to the house, assuming I've got that far . . . so it's becoming a real limitation socially. I

tend to apologize (very British) and explain I have a paralyzed vocal cord and this is now my normal voice, and, no, it doesn't hurt. We then just carry on.

People change toward one and vice versa—I feel it's distracting to them, thinking about my voice and not what sort of person I am, and what I have to say. That makes me withdraw a bit. It's hard to put myself out there anyway with the feeling that people are being kind or sorry for me if they give me their time to chat. I never felt that way before.

Again and again, people talked of their conversations as a myriad of kinds or categories, with varieties of intimacy, size, duration, and stress. Talking with partner or children was fine, but with friends, especially in public places, it was problematic, with larger meetings avoided altogether. It was not just the need to project the voice and overcome background noise; there was also anxiety in front of others, drawing attention to the voice rather than to what is being said. People told of how much easier it was to chat with family and close friends, while more formal groups made them tense and stiff. We all alter our patterns of speech and even sometimes our accents for different groups, tailoring utterances to the social situation.[2] Those with vocal cord problems find this difficult, which can further inhibit their already fragile speech acts.

Several people agreed with Tess and Elaine that their loss of singing was huge. Beata had a vocal cord paralysis at around age fifty, nearly twenty years ago:

In the beginning I found myself boring. How does an illness become an identity? No loud jokes, no imitating of a cabaret artist anymore, and no more singing. The last one was the hardest.

My whole life I sang. As a little child I sang, in the car, on my bike, in bed, with my friends . . . always singing. When I was older, I went to several choirs. In the choir I played the notes with my fingers as if they were the piano, so I could sing unknown songs right away. I was a teacher for twenty-five years and my class was known as the singing class; sometimes I taught by singing.

It was hard to know that was over. Now with new songs I don't know the words, because I don't sing them. Sitting in the car listening to the radio a few years ago I felt there was something different. I realized now I don't sing the songs in my head anymore. My son plays the guitar and sings in a band, so it has passed to the next generation. I am happy with that—no sad feelings anymore.

For Martha, too, singing is a huge loss:

I am far from fabulous—strictly amateur—but I've always sung. In my whole life the only times I've not been in choirs were three years in London and three in

Tokyo. From age nine to forty-seven, six years only with no choir. That is gone, just ripped away. It was a family joke that I had a line of a song for any situation and would break into it. We've lost that. I guess that's part of identity.

It was the fun thing I did with other people, but in addition my emotions were quite tied up with singing. Happy, I sang. Scared, I sang. Anxious, the deep breathing of singing helps. Without being able to sing, it's as if I've lost part of the emotion—the experience and not just the means of expressing or managing it.

Identity

How we are viewed is not simply about looks, facial expressions, clothes, and the like. Those with vocal cord paralysis soon learn that you are seen and judged differently according to one's voice. Some people could not see—or listen—beyond the voice. If one element of identity is off-kilter, then this dominates others. People might see someone's voice problem but not recognize how that affects their whole perception of the person. One person wrote that whereas his voice problems were recognized right away, the effects on his identity effects were not. "How does one become one's disease?" For Martha, the change of voice and identity was "horrible":

> I never realized before how important voice is. I'm very verbal—I think best by discussing and pulling apart a topic. I still do but am now more conscious of it being an effort. I am much more hesitant to speak, and feel I have lost authority, though having been back at work for six months I know my colleagues don't feel that, so it is purely psychological. No one has ever been anything but kind and generous about it, but I hate that I have to consider that they need to be "generous" to accept this weak hoarse and breathy voice.

Making others understand the problem can prove difficult; for instance, many think that if your vocal cords are paralyzed, then you cannot talk. Having a rare and largely unknown condition is isolating. For Meg,

> Importantly, even at a comparatively late age, finding the Facebook Vocal Cord Paralysis group has given me new perspectives. *I am not alone!*
>
> There is a lot of emotion within our collective experience with vocal cord paralysis. We struggle with group activities. Others don't totally realize how difficult it can be to attend a party or wedding reception in a noisy room where we cannot be heard. We sit silently, observing rather than participating. We're there but not totally. Even those close to us don't always see our struggle or our pain around not wanting to go to a party in a noisy room where we cannot be heard or a social situation requiring "small talk." Voice volume, or lack of it, is not really talked about but is a major issue. Others just do not get it. Some of the people

close to us may understand how we feel but continue to express disappointment when we decline invitations.

Some people still think I should just try harder to be sociable, that I should go to brunch in large noisy restaurants where I cannot converse with others and end up just sitting there quietly (like I just don't care) while everyone else is talking and joking and getting to know each other. Why bother when it just makes me feel like a loser? It is too painful! And accomplishes nothing.

For some the worst thing was that people, even doctors, do not understand what vocal cord paralysis means and tend to shrug it away:

> They treat you as if nothing is wrong. You feel misunderstood by doctors, specialists, and even friends and family. We should be treated by doctors who also address the side effects, that is, the psychological, social, and mental consequences of the illness.

One woman, now aged seventy, and twenty years after her vocal cord palsy, echoed this:

> Not going out, especially avoiding noisy places like pubs, theater and concerts, parties, because they will not be able to hear you. You avoid people and people avoid you because they have to ask you several times what you were saying. People are often ignorant about vocal cord palsy and do not understand it. They have no idea what it means to live with. They may treat it as if it were a sore throat. After all, you are not dying, it is only your voice; you have no cancer. After a while you are not invited to parties, to the pub, theater, or concerts.

Another after a lifetime without a voice volunteered, "I have spent years feeling my loss but without having the opportunity to tell anyone, half wanting a pity-party for what my life could have been." Seeking neither pity nor attention, he immediately apologized.

It seems likely that how a person views their palsy is dependent to an extent on its cause. Both Meg and Martha were sanguine. Martha's was the cost of a cure from cancer:

> I didn't die of my cancer—and I don't think I'm going to. The vocal cord paralysis has been the hardest part but I consider myself like Ariel; I gave up my voice for something more important.[3] I suspect this will be a major part of how people handle it: those who got here from a virus or surgery will feel very different from those of us who feel we've dodged a bullet.

Similarly, despite the bilateral vocal cord paralysis affecting her childhood and adult life, "The bottom line is that my medical team gave me the best possible ending while curing my cancer. I survived." Those with palsies of unknown cause or following more minor surgery may harbor less gratitude.

Martha, with a left vocal cord palsy, had her whole esophagus and the top third of her stomach removed for cancer. The rest of her stomach was pulled up and fashioned into a tube that joins her throat. During the operation her left recurrent laryngeal nerve was damaged. Immediately after swallowing, it felt weird; over the next five months scar tissue reduced her swallowing and her newly fashioned esophagus needed dilating eight times. She can drink, but coughs with crumbly things and cannot eat cereal or lumpy soup; sometimes she has to swallow several times to get food down:

> An esophagectomy leaves one with permanent eating changes. I have to eat five times a day, small meals: protein first, gluten free, a complete pain. I have to sleep up on a wedge pillow on my back to avoid aspirating stomach contents. This gives me back and shoulder tension so I need to stretch and do daily yoga—which has to be done first thing because after I eat or drink I can't turn upside since there's nothing to stop the stomach contents coming back up.
>
> But with all this, the absolute worst thing has been the voice loss. Especially when my children (seven, thirteen, and sixteen at the time) have each said, independently, that they can't remember what my voice sounded like before.

We project ourselves into the world in many ways: appearance, dress, makeup, through work and play and through gesture and voice, whether contents of speech and opinions or prosody and emphasis. Vocal cord paralysis reduces how people can reveal themselves and express their emotional experiences. Some even suggested it has altered their interior, emotional lives too.

While difficulties in breathing and swallowing are constant, personal, and mechanical, the voice problems are more pervasive and public, affecting social relations, identity, and how a person views themself. Our appearance is usually a constant, but speech is a performance requiring—for many with vocal cord problems—effort and calculation each day. These narratives of unwanted muteness have most of all shown how vocal cord paralysis can make more difficult that simplest and yet most profound of human needs: to be together, whether to talk, chat, sing, or gossip. In such social glue we reveal ourselves, make friends, cement family ties, and come alive. Other ways of living and other social interactions are possible and rich and rewarding, as we have seen. But just as vocal cord paralysis is hardly known, so the effort and ingenuity required to live with it are beyond most people's awareness or comprehension.

4 We Eat Eels Every Day: Spasmodic Dysphonia

The voice is produced by a complex series of anatomical structures, from lungs to throat, tongue, lips, and nose, which we move automatically with little or no conscious attention. How do I say a word or sentence? How do I make a silly accent? We know what to do without knowing how we do it, with our conscious feedback coming from the sounds that emerge, rather than the movements that underpin them. We don't know how we open or shut or stretch the vocal cords, move the larynx in the throat, or control the palatal muscles. These just happen at our bidding, and we attend to these only when something goes wrong. Normal speech has a volume and a pitch, as spoken phonemes are combined into sentences. But perhaps above all, speech, like song, has prosody and rhythm, with stresses and intonations giving it a melodic, musical, expressive quality.

All this requires exquisite control by the nervous system, involving the brain and feedback from the larynx and airway.[1] As we will see, this can be affected by a number of conditions, like Parkinson's, or after a stroke. More rarely, it can happen on its own, leading to bizarre and poorly recognized problems. The cords may move normally during swallowing, breathing, and even crying, but their coordination during speech is lost, and when either opening or closing the cords, they jam and become uncoordinated, so speech, with its smooth cadences and forward progressions, all but falls apart.

Whenever we practice or over-practice a given movement for years on end, the nerves in the brain coordinating this can become scrambled and no longer are capable of controlling their usual movement. What is curious in these conditions, collectively called "focal action dystonia," is that only a specific movement is involved: for instance, in writer's cramp, only writing is affected, while other movements of the same muscles and the same body part are not. The problem is specific for the action and reflects poor control in the brain by a specific group of nerve cells involved in that one, repeated action. Another example is in sport, where years of training can tip a person beyond their hard-won excellence to coarse, uncontrolled movements known in golf as the yips. Small putts are missed as the intended movement is highjacked by poor neural control. This is thought to be due to abnormalities in neurotransmitters that enable cells in the brain communicate with each other, or in the ways the brain processes information and generates commands to move. Imaging studies have suggested both of these in writer's cramp, where the muscles in the forearm tighten while gripping a pen, making writing impossible. Reduced dopamine release has been reported in the area of the brain known as the basal ganglia striatum together with abnormalities in the circuits involving the premotor and primary motor cortex. Some have also suggested that though the condition is seen as one relating to muscle activation, it may in part be due to faulty or overstretched feedback of movement and position. After all, exquisite control of movement requires exquisite feedback, and this too may breakdown with overuse.[2]

Some classical musicians practice five or six hours per day to maintain their skill, maintaining these neural connections to such precision that it is unsustainable in the long term. Pianists, guitarists, and woodwind players may be more vulnerable to focal action dystonia as a result, with the loss of control being seen during fast passages, leading to irregularity of trills or, for instance, involuntary flexion of one or more fingers in guitarists.[3] Often the wrong muscles are activated as well as the right ones, leading to so-called co-contraction of antagonistic muscle groups. In pianist's cramp, muscles that both flex the wrist and fingers downward and elevate them upward can be activated together, freezing the hand in mid-air. For gymnasts, their whole body becomes their skilled instrument, with not only the movements but their order and the trajectory through the air being overtrained. As Simone Biles showed at the 2020 Olympics (held in 2021 due to COVID), gymnasts dread the "twisties" or "lost move syndrome."

Where localized, these abnormal muscle activations can be reduced by injection of the temporary paralyzing agent Botox, but this does not treat the underlying condition, just its physical manifestation, and it is not always very effective. There is no cure, and the condition is so severe that some athletes and professional musicians have had to retire. Altenmüller and others have suggested that it can be seen in approximately 1 percent of all professional musicians.[4]

When you consider the exquisite neural control and feedback from the cords and vocal tract, it is perhaps surprising that focal action dystonia of the voice, or spasmodic dystonia (SD), is not more common. After all, we talk every day and often for long periods, even those of us who are not horse racing commentators or auctioneers. Possibly the neural systems underpinning speech have evolved to be high-use, so they may be less vulnerable than those underpinning less ecological movements, for example, guitarist's finger. That the condition does reflect overuse and overtraining is suggested by operatic singers' increased vulnerability. The jamming can also occur for some vocal actions but not others, so can be task- and/or pitch-specific, when singing at some pitches but not at others, and not at all for other activities like opening and closing the cords during breathing or swallowing.[5] It can also be either adductor, affecting muscles that close the cords during speech, or abductor, named after the muscle controlling opening of the cords. This is further evidence that the problem is not in the muscles or the cords but in their functional control by nerve cells in the brain.

To imagine the cords in spasmodic dysphonia, it may be useful to consider guitarist's dystonia. The fingers of the left hand (in a right hander) push the strings down onto the frets. In dystonia a given finger may not press down, or may join another finger at the wrong time, or may not lift off properly, whatever the player tries. What usually occurs can intrude into thoughts, even though the wrong movements cannot be controlled. Sequences of movements and their timings break down and cannot be recaptured.

The larynx is involved in many movements to elaborate sounds: it moves up and down in the throat, the cartilages on which the cords sit tilt, and the cords stretch to reach higher notes, and vibrate when together, but their opening and shutting are the most important. It is these that are affected in spasmodic dysphonia. The more common form is a disorder of the cords coming together, or adduction movements. The voice becomes hesitant, strained, slow, and with false starts and stops. Someone kindly sent an audio file of her trying to count, the voice slow, strained and strangled:

h-h-h-eight-y-one—eig—t-two, h-h-h-EIGHT-y—three.

w-w-we-eat—e- eels—EV—every—d—d—day.

Mon-day -tw- twenty—s -s—sixth—uh—APRIL.

Each word was recognizable, but the cadence and melody were absent and her speech lacked progression and timing, which would render to-and-fro conversation all but impossible. Underpinning this, the cords would be stuck together, making speech clunky and cramped. "We eat eels every day" is a stock phrase to assist the diagnosis of adductor SD, since vowels are made with open cords, something most difficult in this condition, whereas for abductor SD, in which the difficulty is in opening, a stock phrase is "Harry had a heavy heart." Tragic though these professional musicians' and athletes' conditions are, they are specific to playing their instrument or to playing their chosen game. In contrast, the people we now turn to with vocal spasmodic dysphonia find it present during their normal, everyday speech, with life-changing consequences.

Say "Uganda"

Jude read my inquiry on a support group and replied immediately, with Kafkaesque opening:

> In 2008, aged thirty-six, after a year of what I would describe as quite a few traumatic experiences—miscarriage, divorce, and physical assault—I woke one day with what I thought was a bad case of laryngitis.

Suddenly, she felt as though someone had a hand around her throat as her voice was strangled, strange and interrupted. Unlike Gregor Samsa, she had no idea what had happened.

There followed two years of doctors' visits, including many specialists, before they diagnosed a small polyp on her vocal cord, which was then removed. It made no difference. For twelve months, twice a week, she drove fifty miles or so to a vocal coach and in between would religiously do her vocal exercises. This had little effect on her broken speech, so the coach suggested returning to the ENT surgeon. Once more, he could find nothing wrong, so referred her to a psychiatrist, suggesting her problem might be "emotional." By this point she certainly was emotional, questioning her own sanity. Surely something was wrong, but then why did the ENT not find it? Questioning herself over and over, she thought maybe it was because of all she had been through in such a short time.

The psychiatrist saw her once and discharged her as being a "well-balanced, emotionally healthy woman." She felt relieved but did not know where to go next. Physical medicine had found nothing, and psychiatry pronounced her completely sane. Abandoned by both, she was on her own with a severe problem without a name, explanation, or treatment, and which she could not explain to those around her:

> Frustration was my overriding feeling. I had exhausted my options since my ENT surgeon, vocal therapist, and psychiatrist had no answers. I felt that I would just have to get on with it and cope as best I could. I could not make sense of it and was just tired by this point of being asked all the time, "What is wrong with your voice?"

Even her boss, who had been incredibly patient, began to get annoyed and frustrated, which affected Jude's confidence terribly. She was a sales director for a large cosmetics company. Joining straight from school and working for them for over a decade, she was well regarded and had always prided herself on being good at her job. Unfortunately it involved public speaking and presenting, and this became embarrassing. If she had a presentation coming up, she would conserve her voice for a few days beforehand, as best she could. Frequently, it was her opinion that was needed. Whereas before she would argue her case, now she found herself choosing her battles by how much voice she could muster; other times she just rolled with it.

She was very open, explaining that she had a problem with her voice that had not been identified yet, though sometimes it was easier to say it was laryngitis, and explain she was not in discomfort or pain nor contagious. Her main strategy was to avoid large gatherings and new people. She'd email rather than have a telephone conversation. In sales, building rapport was important, and she tried for face-to-face meetings rather than use a phone. Quite often, if a call came up on her mobile she would ignore it; if the person left a message she would text back. She thought that if she had not been so respected her company might have let her go, and hated to think she was being carried.

She tried every possible alternative remedy: yoga, meditation, cranial sacral therapy, hypnotherapy, various adjustments to her diet, and cutting out caffeine. Nothing had the slightest effect. It was becoming increasingly exhausting to speak, and she began to avoid social situations and phone calls to save what little voice she had for work; in doing so she became increasingly isolated—even her mother phoned less.

Several years later, her sixteen-year-old nephew phoned to tell her to turn on the TV: there was a woman who sounded just like her, talking about spasmodic dysphonia. Jude tuned in and was so excited: she *was* just like that. She looked it up online and found a specialist at a local university hospital. She was delighted but also nervous, in case he said she did not have it. The specialist listened, asked her to say sentences rich in vowels (said with open cords), scoped her, and agreed she had spasmodic dysphonia. "The relief was amazing. I was not going mad!"

Next was the "emotional rollercoaster" of Botox. For three years or so she would go every twelve or fifteen weeks and have it injected through the anterior throat under electromyography control into an adductor muscle. Despite no local anesthetic she did not find it painful or uncomfortable. The results, however, were hit or miss. Sometimes it was effective only for a couple of weeks, but often it was less successful. The sense of expectation and not knowing how good it would be were difficult to endure. She would plan her diary around injections, arranging big presentations when she hoped she'd be at her best.

Despite all this, divorced and single again she wanted to meet someone and start a family. A friend kept talking about a man she knew, but it took Jude a full two years to agree to meet him, and then only after the friend had explained Jude's predicament to him. For their early dates, if it was somewhere loud she would write down what she wanted to say on Post-it pads. Soon they would meet for walks and sought out quiet places. Even then she was hesitant, her voice "a husky whisper; I felt so ugly speaking." She remembers at one stage phoning a beauty salon for a leg wax. When she arrived, the beautician said she was expecting an eighty-year-old lady from her voice. Jude was so embarrassed she never booked that beautician again. Generally, she'd explain her problem; otherwise, her anxiety waiting for them to raise it overshadowed everything else and was mentally exhausting.

She and her date married and had a child. Jude then had to adapt parenting to her voice. Reading bedtime stories was challenging, so they listened to a lot of audio stories together. Her daughter soon understood that her mother could not speak loudly and knew to stay close to Jude when on her scooter because "Mummy can't shout if I go too close to the road." As her daughter grew—she is now nine—she would speak on her mother's behalf, especially if someone was asking for a postcode or telephone number. She would say, "My mummy has a poorly voice so I will tell you."

Despite this, dysphonia still dominated her life. It was exhausting to balance family life and work. She had so much in her head but was rationed in what she could say. At one stage she felt she had let her husband and daughter down and that she could not be a proper mother or wife. At work she also felt a burden and she began to have dark, suicidal thoughts. Her general practitioner (GP) prescribed sleeping tablets and amitriptyline, while Jude sought other, more natural ways to improve her mood; fortunately, this time passed. This was a decade or so into SD, during which time she had married, had her much-loved daughter, and was working—from the outside all measures of success. That this happened so late shows how pervasive and enduring she found the continuing effort to live with her problem.

She endured nearly twenty years of this before she saw a video of a woman from South Africa who had gone to Japan for an operation, a type 2 thyroplasty. Since the vocal cords close too tightly in adductor SD, the surgeon separates them a little, and then inserts a small titanium bridge at the base of the cords to maintain the separation. It is usually done under local anesthesia with sedation so that the patient can talk while the surgeon adjusts the vocal cords to the optimal distance for the voice. This leaves the cords slightly further apart so the voice is breathier but less strained.

Jude researched it extensively online and looked at the Japanese surgical team's results, talked it over with her husband and her mum, and decided to go for it. After Skype calls and having sent multiple voice samples, she was accepted. Concerned not to have the pressure of a return-to-work date hanging over her rehabilitation, she left her job.

One big drawback was traveling to and from Japan on her own. She used to hate dropping off her daughter at school; all the mums gathered for a chat, which she found difficult, and she knew they thought her snooty. But one morning she was chatting to another mother and explained she was going for an operation in Japan. Jude explained she was going alone, and the mother offered to go with her. Japan was on her bucket list and anyway she couldn't think of Jude going alone, even though she was not a close friend. Jude dreaded the thought of travel, particularly the return from Japan when she would have no voice, so was delighted.

The team at the hospital was very honest and gave her a fifty/fifty chance of success. She had done her research rigorously and trusted the team. "The surgery was the single most surreal moment of my life, lying on a hospital bed in Japan with my throat open, speaking while they adjusted the bridge." The head surgeon spoke good English, but the others knew little. Two nurses, with no English, supported her by massaging her feet.

It was also surreal because the word I had to keep repeating during surgery was "Uganda," since vowels are particularly difficult when you have adductor SD. When they first put the bridge in and I said, "Uganda," my voice was so airy and whispery, so they adjusted it and I said, "Uganda" again and it was really deep and almost sounded like a horror film kind of voice. They made the third adjustment—the surgeon said they were "fine tuning her voice like a piano," and I finally said, "Uganda" and heard the closest thing to my old voice I had heard for a very long time. It was an overwhelming rush of relief and gratitude that this surgery could work for me. The possibility that I might have found some resolution just filled me with hope and joy.

Lying on the operating table, she felt tears rolling down her face. For the journey back, she carried a lanyard round her neck that read: "Post vocal surgery, I'm sorry I am unable to talk."

Many people think I'm incredibly brave to do what I did but if I'm really honest I didn't feel I could live that way anymore and was desperate to try anything. I was just fortunate. The moment I heard my "old" voice is one I will never ever forget. I feel like I have been given a new chance at a better life and will forever be grateful to the surgical team in Japan. I have been fortunate enough to have the financial support of my mum, who enabled me to go for the operation, and of my husband, who supported me quitting my well-paid job to enable a stress-free recovery from the op.

Back home she spent six months recovering. The result of surgery was miraculous. Her daughter had never heard her voice and said, "Mummy, your voice is so pretty." Whereas before talking involved strain and planning, afterward it was effortless once more. She knows it is not her original voice, but surgery has been life-changing nonetheless. Her voice is breathy; her brother talks of it "as being like Marilyn" (Monroe), which she can live with. Having left her job, she took her time, and once she knew her voice was strong enough, she retrained as a hypnotherapist. The other students all agreed how great her voice was for her new profession, which involved her talking for an hour with up to seven clients per day, helping them with stress. This has become her love and her passion. She is even going to Sarajevo to help those affected by counseling war victims.

I'm under no illusions that I am "fixed"; I will always have SD, but the bridge enables me to function. My voice is still very husky and soft but I have my life back. SD is so much more than a vocal problem. Over those years my self-esteem, confidence, mental health, and everything—work, family, friends, identity—was affected.

We exchanged emails and then met on Zoom. We spoke for over an hour, and then she had more clients to see, her daughter to pick up from school, dinner to prepare, a husband to catch up with; in short, she had her life back. She would not have chosen SD and remembers the long years where it dominated her days and remains frustrated that so many in the SD support group remain stuck. Her surgery remains controversial and available in only a few centers around the world. She would like to change that. "SD is not a life-threatening condition, but the surgery is a life-changing operation."

Boxes

Not everyone with adductor SD travels to Japan. Gwen, like Jude, has lived with it for over twenty years. She worked in a pathology lab as a bacteriology technician. After having children she moved to a large supermarket in its abattoir, swabbing surfaces and checking cleanliness. "The pigs came in underneath, and we got off-cuts of meat." Her husband was in the Air Force, so they moved around. When he retired, they started a delicatessen, but after a few years they divorced. Next she was a doctor's secretary, and then an estate agent. It was better money, and she'd always loved other people's homes.

Her passion in life from age five has been acting and singing. With only a slight pause when her two boys were young, she has been on amateur stages with leads in plays and musicals and solos in choirs for sixty years. Now in her late seventies, pin-sharp and interested, she reeled off musicals: *Guys and Dolls* (twice), *Oliver* (twice too as Nancy), *The King and I, Sound of Music, Oklahoma, Brigadoon, 42nd Street.* There were too many plays to remember, from light comedy to J. B. Priestley. In her fifties, as the leads dried up, she graduated to directing, "almost as good as being on stage." She directed plays and musicals, including one with tap dancing, and loved it all.

She also sang in amateur choirs and there met her present partner, Don. Ten years older than her and now slightly deaf, he was a trumpeter till his front teeth dropped out and so retrained on the tenor horn, which allows a more forgiving embouchure.

Her voice problems began over twenty years ago, with difficulty moving from high notes to low. She assumed she had strained her vocal cords and carried on. Gradually, it worsened and her speaking voice became gravelly. She could still direct shows, so took little notice.

Seven years ago she and Don retired to a small town in rural Dorset, chosen for its active arts center for her theatricals. Once moved in she went to an ENT surgeon, who immediately diagnosed adductor spasmodic dysphonia. Gwen was fascinated by her voice analysis and the scoping of her larynx. "I saw how my vocal cords spasmed together when I talked," she noted, and was delighted to have a diagnosis after all those years. But it still did not seem bad enough to treat.

She went back to the surgeon when she could manage only two or three words before her voice seized up and when singing became impossible. At home with Don's poor hearing, she had to keep repeating herself, and it became "easier not to talk."

She agreed to Botox, which would allow the cords to rest relaxed in the midline, but with sufficient movement to improve her voice and protect her airway during swallowing. In the United Kingdom, healthcare is funded by comparatively small groups of primary care physicians known as Clinical Commissioning Groups (CCGs). They decide what they fund in the National Health Service. Despite neighboring areas providing Botox for free, and her own allowing its use for many other conditions, her CCG decided not to pay, so she bought the drug herself. Though she needed only two or so units, it was only sold in bottles of fifty units, costing hundreds of pounds. Gwen bit the bullet and bought the Botox.

The outpatient procedure was entirely trouble-free. She sat on a chair like a dentist's. Her surgeon, Miss J., introduced a needle through her larynx, found the right muscles, and injected. Gwen's voice gradually improved over the next four weeks or so, and this lasted for three to four months. During that time she talked and talked, to Don, to friends, to anyone, making up for lost time. She rejoined the choir, curious that week by week she could reach higher and higher notes as the bottom notes disappeared. As a result, she changed her position in the choir as her vocal range moved. Then COVID-19 came along, so the choir did not meet. She found other challenges. Once retired, she learned clarinet and post-COVID performed in the town orchestra. Even with the exquisite control of breathing needed for a woodwind instrument, she has never had a problem.

She sent a video of before and after Botox to the CCG. They sent it back with a letter, thanking her but saying there would no change in funding. She does not even know if they looked at the video. She had self-funded another Botox injection just before the COVID pandemic, but then for nearly two years, throughout the pandemic as the hospital stopped seeing routine outpatients, she and Don sat tight, isolated by the pandemic and by her increasingly tight voice. She worried that her surgeon, Miss J., might be ill. "I don't want to go anywhere else. I trust her."

Eventually, Miss J. called, and an appointment was made. She had found two more patients who needed Botox for the same condition, so they clubbed together to buy a bottle of Botox. Though the injection was a bit more traumatic (maybe Miss J. had become "a little ring rusty" over the fallow period), afterward her voice was back. As the three patients waited their turn they had a great time chatting; after the eighteen-month isolation it was their first chance to compare notes—and voices.

Without Botox, Gwen readily admits, she could have just about carried on, more of a recluse and talking little at home. Her beloved lunch group of four "girls" would still have met and she would have laughed with them and eaten with them, but hardly spoken. How much better it is just to go and chat about everything and nothing, without strain or effort. Even now each morning she wakes and wonders what her voice will be like, and at times she even worries she may have "forgotten how to talk normally." Her voice is not normal, but it will do and is so much better than at its worst.

Stage and choir are memories. She has a large box of old programs from her past productions, congratulatory letters and photographs, and all her old scripts, which she can't bear to throw away. Looking through them is upsetting, but also makes her proud. In one play her character always finished others' sentences. Those days are gone, but with periodic Botox and her own reserved resilience, she will find a way. She tries not to think about what she is missing by keeping her emotions deep down. If they surface, she is still raw and "cross" at not being able to do what she so loves. So, she hides them away, in her own invisible box and like a proper hoofer gets on with life.

Though there are at least two muscles that close the vocal cords on either side, there is only one, posterior cricoarytenoid (PCA), that opens them. Of course, opening is crucial for breathing as well as for speech. This action can also be affected in abductor spasmodic dysphonia. It also is more difficult to alleviate with Botox since one can inject only one side a time, to avoid risking paralysis of both cord opening muscles (which would be called in medicine a "never event" for obvious reasons: there would be no muscle to open the cords). It is to a woman with this condition that we now turn.

5   Conversing with Myself: Spasmodic Dysphonia

In 2015, Claire noticed gruffness and an occasional break in her speech. An ENT surgeon scoped her larynx and diagnosed reflux from the stomach. A year later and no better, she went back, only to be told the same: there was nothing wrong with her vocal cords. Things deteriorated, so she asked to see a speech therapist. For six weeks she did the prescribed voice exercises. She was a teacher, and by now it was affecting her work. At some meetings she dared not to speak for fear of how it would come out. Embarrassed, almost ashamed, by the sound of her voice, she avoided social situations:

> I believed I was suffering from extreme anxiety/stress since this is how I sounded. Voice plays a large part in how we perceive ourselves, and since my voice sounded broken and exhausted, this was the message I was receiving from myself.

Against that, other than her voice, her life was actually quite good: she was not feeling stressed and had up until that point been a very sociable person. She certainly didn't feel like she was having a nervous breakdown.

Claire's voice finally gave way in July 2019, during the summer break from teaching, Maybe, she thought, resting her voice would help. But when traveling to London for her son's graduation, she found she could not speak at all. The more she tried, the worse the spasms in her vocal cords became:

> My voice was all over the place and, as a result, so was I. It was truly awful. I could no longer pretend; in that moment I just let it go and stopped trying to sound normal.

She went back to the ENT surgeon, without much success. In the end and after much research, it was Claire herself who suggested spasmodic dysphonia. The surgeon agreed to refer her to a specialist, admitting he had never seen it. The specialist agreed and diagnosed the rarer and more difficult form of spasmodic dysphonia, abductor, in which there is spasm of the muscle moving the cords apart rather than bringing them together.

She was told the only treatment was Botox injections to the PCA muscle on one side. Since Botox paralyzes the muscle for several months, injections are never given to both sides because then the patient's cords would be closed. Her local consultant suggested the first under a general anesthetic. She had four units of Botox injected into her left PCA muscle. She woke up and waited. Her voice was a bit deeper, but the spasms were as present as before (suggesting the site of injection was inaccurate). This injection was just before COVID hit, and after that most routine services closed and she was left high and dry.

Claire taught philosophy and religion at an independent boarding school, which moved to online teaching with COVID to ensure their fees were paid. Without effective treatment, though, she was voiceless. Realizing her livelihood was at risk and knowing that the local surgeon appeared to lack the necessary expertise to locate the right muscle, she researched consultants who specialized in Botox injection and contacted the top three. The most impressive and first to reply was a Harley Street surgeon.

He was incredibly helpful. The treatment was private, and so she paid, despite knowing that many other patients at the clinic were having their treatment paid for through the National Health Service (NHS). Throughout the country, some people would be allowed Botox and others denied it, determined not by clinical need but by local committees (CCGs) balancing budgets across diseases, though of course politicians deny there is a "postcode lottery."

Claire's husband was a professional musician, and with COVID lockdown all his work was canceled indefinitely. They were surviving on her wages alone, so they had to borrow the money for the injection from a family member. Without that, her job was over and their income gone. Fortunately, the injection was a success, with a significant improvement in her speech, which was such a relief. She knew the injection would last only a few months, but despite her best efforts she could not find an NHS department treating patients, so she returned to Harley Street. Once more she applied to her CCG, confident of success this time, since she could show that she was unable to find the treatment that had been so successful anywhere else. A few days before the next injection, the CCG notified her that they would not fund the treatment as it was too expensive. Claire was furious. Given that the alternative was that she would lose her job and then two people would have to rely on social service support, this was financially short-sighted as well as clinically unusual and morally question-able. "The CCG had little or no understanding of the impact or treatment of this disorder."

Their financial situation had worsened as the pandemic took hold, so they decided to fund the treatment with a credit card. They traveled to London on deserted roads and parked outside the Harley Street clinic, the usual displays of parked Mercedes, BMWs, and Teslas nowhere to be seen. Once more the injection went well. When she went to pay, they told her it was free; the surgeon and his staff were all so shocked at how she been treated by the CCG. She broke down, quite overwhelmed. Curiously, her crying voice is almost perfect:

> This was the first time that anyone had acknowledged the difficulties I was expe-riencing. I was speechless at this act of kindness.

Back home she appealed to the CCG several times, supported by her local MP, but the CCG refused to budge.

She had to wait until the end of the lockdown before she was considered for treatment again. Lockdown was an isolating period for many people, but without a voice, wearing a mask, and having a two-meter distance most of the time made it impossible to have any type of meaningful communica-tion. Like many others, she struggled through this period as best she could.

Unable to continue going to Harley Street forever, she explored other NHS options. Her original local clinic said they could now inject her with Botox. However, the way they spoke about it suggested they had little or no experience of the procedure. She knew the problems that could arise from injecting too much Botox or missing the PCA muscle, but desperate for some relief, she agreed.

At the appointment her fears were confirmed. She found the whole thing truly horrific and at times believed she would die.

There were two consultants, one guided by the other, together with a neurophysiologist whose job was to record muscle activity to make sure they were in the right place. The consultant administered two sprays of local anesthetic up each nostril and then proceeded with the endoscopy, a small flexible tube up the nose, to see which vocal cord the last Botox injection had been administered to and which side looked the weakest. Then she sprayed local anesthetic to the back of her throat. Claire began to choke and wheeze and found it difficult to breathe. This calmed down and then more spray was applied to the back of her throat. Again, she began to wheeze and feel as though she couldn't breathe. This may not have been because of any incorrect procedure or reaction. We feel air passing over the lining of the larynx and upper respiratory tract as a way of gauging that we are sucking air in and out. Once the anesthetic works, we no longer feel this, which can make us think we are not passing air back and forth at all.

They suggested she breathe through her nose to regain normal breathing and reduce the laryngeal spasm. She said that she was finding the procedure too traumatic to continue. They reassured her she was safe and that they now needed to inject local anesthetic through the front of her throat. Almost immediately her airway closed, and she was unable to breathe either through her nose or mouth. She began to panic and feel dizzy. The wheezing continued and they agreed to stop to procedure. In any case, she did not feel safe enough to continue. The consultants examined her neck, with one instructing the other where the correct entry point was so she would know for next time. It became clear that her consultant had not previously done this, which made Claire feel incredibly unsafe. They also suggested she had a hypersensitive airway.

She left very distressed, unable to swallow due to the local anesthetic (without sensation she risked choking), and sat on her own in the car park, hardly daring to move in case her airway closed again. She had never felt so terrified in her life. She again appealed to the CCG and explained that the consultant had no experience administering this procedure and that she felt unsafe. They spoke with the consultant, who said she had treated "less than three people" but would not reveal the true number for reasons of "patient confidentiality." Claire did not return.

Instead, she approached the second on her original ENT list who was an expert in abductor spasmodic dysphonia. She was reassured by both his knowledge and his manner, and he was an NHS consultant, with his own funding. She went for two injections with some relief, but unfortunately she had an adverse reaction to both, with laryngospasm. The second was so severe they had to call an ambulance. Her airway was almost shut, and she was terrified. Because of the pandemic, she went in the ambulance to hospital alone. Wearing a mask, barely able to breathe, and certainly unable to speak, she sat in a wheelchair in the emergency department holding up a note attempting to explain her condition. The consultant in the emergency department apologized many times that he had no knowledge of the condition, but was very kind and sympathetic, and treated her as having had an allergic reaction. Slowly, she improved.

This episode worried her hugely, as she realized the full impact of losing her voice and being incapable of explaining or indicating what was wrong. "It was devastating. This did not just affect me, but my whole family. How could I possibly explain if my child was ill, or needed help?" After this episode they decided not to proceed with any further injections.[1] Her voice remained so severely affected that despite all her best efforts within and beyond NHS medicine, her job was lost:

> When I think of what has happened in a rational manner, losing your voice doesn't sound so bad. . . . However, this is not the case. I have lost a great deal. At the end of next month, I am officially unemployed, I have lost the job that I loved. There isn't much I can say other than I am unable to teach and do what I once did. I now am completely lost. All my qualifications and work have been in this area, and I now have the daunting prospect at almost fifty years old and with no voice of attempting to find work.

On paper she is very impressive: a successful teacher and head of department for many years, a degree in English literature, a postgraduate certificate in education, and an MA from Oxford. She is also completing an MSc in Oxford as well. However, trying to get prospective employers to see past her vocal problem is virtually impossible:

> Previously, I have never been unemployed, ever. I've had interviews for crap jobs; I can't even get a job at McDonalds. Most jobs require interaction and so require speech.

She was teaching online once and a boy asked lots of questions she could not answer, which annoyed her. Claire can no longer think about the complexities of what she is teaching and discussing and think about speaking as an act at the same time. There is just not the necessary level of connection between teacher and pupil. Indeed, thinking is what she seems limited in too, rather than expressing:

> My thoughts are sometimes deafening, I believe more so now than ever, a consequence of being unable to express them vocally. It's difficult to explain, but my internal dialogue remains fluent and incredibly eloquent. The difficulty here is that when I try to vocalize my thoughts, what I produce does not match. My speech is labored, slow, and difficult. I am so focused on the physical action of speech that I limit the words I use, limit what I say; sometimes I can't even be bothered any more. That is one reason writing is therapeutic.

There is such loneliness since the only person I can authentically converse with is myself. I struggle in most social situations as my speech is unable to interact, interject, or contribute to conversations, but my internal dialogue still does. It's quite a claustrophobic experience as I feel trapped inside my thoughts. This limits me as I avoid social situations as I simply find them unpleasant and difficult to deal with. If, for example, I have to sit down and eat with people I don't know too well, I have to deal with eating, drinking, and speaking all in one go and this can be really distressing. If I'm in a situation with people who speak a lot— I'm sure you know the type—I end up sitting there as a passive observer offering a platform to facilitate their monologue. I don't like this, and it seems to happen quite a lot. It has changed my behavior and I don't have much choice. Prior to this I loved socializing, speaking, and laughing with anyone really. As a teacher of religion and philosophy, there was nothing more I loved than debating, discussing, disagreeing on all things related, and for me it brought an immense joy. This has now gone, and I miss it dearly.

But the biggest impact has been on relationships. I was heartbroken for a long time as I was unable to speak properly with my children. We have five children, and they are older but there is still a need as a mother to speak, console, help, and laugh with, and again I am unable to do this. My conversations are jarred and difficult. Also, with my husband I feel as if I'm a different person. I don't really go much further in thinking about this one, as it truly devastates me. I often think of the Ship of Theseus . . . if all these different parts of me have changed, been replaced with something different, and so significantly, am I still the same person? I'm not sure I am. This is by far the most difficult area to deal with.

I sometimes listen to recordings of myself when I could speak, and I sound and look like a different person. I still have not get used to how I sound. At times it is humiliating. I'm usually light and cheerful. . . . I also realize that these are my perceived thoughts surrounding the situation and aren't necessarily true. I'm working on quietening these thoughts as they can be deafening, and when they are negative thoughts it's difficult to contain them. In fact, I'm working on many things at the moment, trying to rebuild my life after the devastation this has caused. I have doubted myself and, in the past, become paranoid about my voice. I've been rock bottom and now do not hide it and sort of have accepted it. Trying to rebuild in a new way that accommodates the changes I have experienced. It's difficult.

We met online via Zoom. I expected a tough time as I tuned into her voice and discussed her searing narrative. Instead, she was engaged and eloquent, the sort of teacher I would have loved as a boy. It seemed extraordinary that she had met so many barriers when all that was needed was a little more time to listen as she found the right words. I soon left her vocal problem behind and just enjoyed listening and conversing. Though not unaware of her impairment, it was neither insuperable nor the main part of how she spoke or who she was. Things just took a little more time. Her personality, her selfhood, was evident in her manner, her gesture, her facial expression, *and* her voice. Her being was not, could not be, limited by the vocal spasms. I realized that our short interview would no doubt be important for her, and so her performance might have been at the top of her game, but, even so, it was difficult to reconcile with all that she had experienced. Had people given her more time and listened more patiently and carefully, then her experiences could have been so different. Her job might not have been jeopardized and her self-confidence so tested.

In the United Kingdom, the Disability Discrimination Act of 1995 made it unlawful to discriminate against disabled persons in connection with employment. A disabled person is defined as "someone with a physical or mental impairment which has a substantial and long-term adverse effect on his ability to conduct normal day-to-day activities." Discrimination was in turn defined as when "a person treats someone with a disability less favorably than he treats or would treat others to whom that reason does not or would not apply and cannot show that the treatment in question is justified and [when] he fails to comply with a duty to make reasonable adjustments imposed on him in relation to the disabled person."[2] The rules are there, but enforcement is a long way off. Of course, the law can sit above and ahead of society as a whole, while pointing it in a direction. But all Claire and countless others need is more time, some understanding, and to feel that others might care.

Oxybate Date

Restless and somewhat desperate, Claire scanned the web looking for treatments and came across a trial in Boston with the drug sodium oxybate, a central nervous system depressant. Noticing that alcohol had improved some people, the doctors, led by Kristina Simonyan, had started using this, which otherwise is used to treat narcolepsy and cataplexy. In the former, people fall sleep uncontrollably, and in the latter, they have a sudden loss of muscle strength and thus often fall after a strong emotion or, equally curiously, laughing.[3]

Claire contacted them and they agreed to enroll her in the trial. She raised the money online, through a gifting site, and went with her husband to the United States. COVID meant she had to quarantine in Canada for two weeks beforehand. This might have been an enjoyable break but for one thing: Claire had to be caffeine- and alcohol-free for a week. It was, she said with feeling, "hell."

The trial took three days, and began with voice recordings, cognitive tests, and an MRI during which she had to speak. Then she was given 50 milliliters of neat vodka and the recordings and cognitive tests repeated. These were repeated after four big shots of vodka, when they also repeated the MRI. By then she was feeling more than a little drunk. Day two was similar but she had either the drug or a placebo in 100 milliliters of fluid, and they took her blood pressure and heart rate before doing the similar tests as before. Day three was the switch to either placebo or drug. The last day she was sure she had the drug: she felt dizzy, needed to empty her bowels, and then after several more minutes felt alert and lucid. More importantly, she felt her voice spasms reduce and her voice deepen. For the rest of that day, miraculously it seemed, spasms were gone: she could talk and so she did, chatting away nonstop to her physicians, to her husband, and even to strangers as she made her way back to her hotel. The next day the SD returned.

A problem is that a bottle of oxybate (180 milliliters) costs $5,000. She heard that there is a CEO who appears on TED Talks with SD, and that he has a swig of his oxybate before he goes on. Once settled back in the United Kingdom, she was onto her hospital requesting the drug on the NHS. She was surprised not to be dismissed and is pursuing it not just on her behalf, but for many others she knows of:

> Remember when I said how SD affected my personality? Well, when my voice came back, even for that brief time, so did I. I am limited, so limited by SD, having the think about speech all the time, and not just exist.

Unfortunately for her, in the United Kingdom medicines are now licensed by a government agency that has not approved oxybate for her condition. There is simply not enough evidence to prescribe yet.[4] So ever-resourceful Claire tried another idea.

Professor Simonyan in Boston is also investigating the use of virtual reality and real-time visual feedback of speech therapeutically. The research team has found that whispering sounds to people with spasmodic dysphonia leads to more normal speech. Claire recorded 200 phrases during whispered (more normal) and usual volume speech (dysphonic), and a computer distinguished between the two. Then, she was given a VR headset to place her in novel and relaxing situations while it gave her phrases to repeat. A visible bar in the VR headset fed back how normal each phrase sounded. She was also encouraged to imagine being in relaxing situations, jogging or doing yoga, or playing guitar. These all seemed to lead to less strangled sounds. There was also a carry-over effect, so each evening she sounded better too. Claire was encouraged, though inevitably the questions are whether a system can be developed for home use and whether it works in the long term.

Like George Eliot, Claire needs to utter in order to be. Equally though, if anyone can find a way to escape from or at least find a way of living with her SD, it is she.

Spasmodic dysphonia is a distinctive and singular sound, but the condition is so rare that few ENT surgeons or speech and language therapists, let alone general practitioners, will recognize it. Yes, there is lots of information online about the condition, but these websites give no clue how difficult it can be to make the diagnosis in the first place when unfamiliar with its symptoms. On one support group site, people shared the delays between onset and diagnosis. It averaged two to four years and much depended, as in the case of Jude and Gwen, on how actively the person pursued matters. One woman saw fifty-nine doctors before her diagnosis. She found Botox less helpful and lost her job too, teaching pupils with learning difficulties and speech problems. A classically trained singer first noticed his voice deteriorating for several years before proper symptoms appeared, which had been present for four years before diagnosis. It was thought to be a post-nasal drip, and then he was offered an operation on his nose cartilage, which he declined. After "loads of speech therapy and lots of cameras" (direct observation of the vocal cords with a flexible laryngoscope), a specialist finally made the diagnosis. Unfortunately, this was after he had left singing and retrained as a teacher, which position has itself become threatened by the condition.[5]

Happy endings, like Jude's, seem rare. It would probably help if undiagnosed speech problems were referred and discussed more readily between centers and doctors, rather than doctors excluding the conditions they know about and leaving it there. With video and even medical tests now routinely recorded, this should not be too difficult. The system surely cannot depend on the enterprise of patients, some of whom have had to make the diagnosis themselves, not least because one is unlikely to meet many as persistent as Jude or Claire. Both have been across the world searching for their voice, with mixed result as of yet. For Claire, the various experiments have allowed one important insight worth all the travel and effort. She now knows that her voice is still intact, in there somewhere, and that she has accessed it, albeit fleetingly. Her task remains to try to release it more consistently.

6 Stuff Happens: Cleft Lip and Palate

How a single fertilized cell develops into a formed, recognizable fetus within the womb remains largely unknown. One of the key parts involves gastrulation, the curling over of the relatively flat embryo onto itself to become tube-shaped and so form the gut. This occurs around weeks three to four post-conception and requires the fusion of the meeting parts in the midline. It is a little later, weeks five to eight, that the lips and then palate meet and fuse. In around one in 700 births this does not occur normally: they do not meet, resulting in a cleft, either between the lips and/or between the soft and occasionally hard palates.

If not repaired through surgery, the child will have difficulties in feeding, hearing, and speech. Nowadays, in the vast majority of cases, surgery is very successful and most babies born with a cleft lip or palate will lead totally normal lives (at least in high-income countries).[1] Children with a cleft lip or palate will be monitored closely as they grow and develop. This will include their hearing, jaw and teeth development, and speech. Most children are able to develop good speech and language skills following surgery. But sometimes a repaired cleft palate can make it difficult for a child to pronounce some sounds clearly. This may affect speech quality, making it sound a bit "nasal." A speech and language therapist will be able to help.

Perhaps the other main problem for those born with a cleft comes from others' response to the condition and whether their culture tolerates teasing and other such behavior. Children may be subject to prejudice because of their appearance and speech, which can lead to emotional and behavioral problems.

When I approached support groups for those adults with cleft, I anticipated some experiences due to their appearance as well as wanting to focus on their speech and its effects, but the resulting narratives were not as expected. These may not be typical, but their importance is they may be generalizable to others with visible or aural difference.[2] I have also written about experiences of those in Europe and the United States; those in different cultures, in low and middle-income countries, may well have led to very different and sometimes harrowing narratives.

Jane

Jane has had more than twenty operations, including three bone grafts to her hard palate, dental implants, lip closure, palate closure, reshaping the nose/lip, skin grafts, and an operation to lengthen the nose to upper lip area—all part of living with bilateral cleft lip and palate. Her father did not know how to cope. Panicked and frightened, he left all the medical stuff to her mum. But this did not stop him from sexually abusing her as a five- to six-year-old child. Her mother played down the cleft, and in doing so reduced its significance beyond what Jane felt appropriate. She wanted it to go away and be fixed medically, which made Jane feel inadequate. It cannot be easy for any parent in this situation. Torn between supporting and smothering her vulnerable child, her mother became overly protective and verbally abusive, "like 'Mommie Dearest.'" Her controlling and criticism compounded Jane's sense of shame.

She would often criticize her daughter's speech, wanting it to go away or be cured:

> I always felt like nobody understood me, and it didn't help that my mother was extremely critical of my speech. She would often try to correct me, or criticize me that when I spoke too fast, she couldn't understand me.

A problem not unique to those with cleft is that she felt her inner voice was fine and perfectly intelligible, and as a result she was not aware how difficult her enunciation was to follow. One of the joys of parenting is to see the acquisition of language in one's child; initially words, then phrases and sentences, to be followed by at times an unfolding and continuous stream of consciousness.[3] Jane was acquiring language without sharing, unable to recognize how difficult her speech was to follow. A medical report from school when she was age four states: "Almost completely unintelligible speech and very little consonant articulation at all. Extremely hyper-nasal with a good deal of nasal airflow which is to be expected." Of course, it was not only language. She did not like to eat in front of others, let alone drink liquid since it would come down her nose.

Her wider family was no more supportive. She gave her grandfather an ash tray with her photo embedded in it and he mocked her. She coped as best she could, by denial:

> When I was young, I pretended I was normal. I avoided all recordings of my voice. In my head, in my thoughts, my voice sounded okay. Speech and language therapy I used to hate. My class made fun of me.

As a result of her prolonged therapy, she remembers learning to speak:

> I remember constant, nonstop speech therapy, practicing where my tongue went and my lips. I always remember being ashamed of hearing my own voice. I sounded too nasal, I couldn't properly pronounce "s" and "t." At one point, I became ashamed of speech therapy and couldn't wait for it to finish.

At age fourteen she had a huge row with her parents about it and refused to go any more:

> I could never blow a balloon up and would have difficulty with eating utensils because my palate was soft or delicate and I often had bandaged or sore lips. These all compounded my shame.

Most of her class ignored her and she was locked out of all groups, with few friends in either elementary or high school. "High school was hell; my only friend was another outcast." Growing up she always related to older people, whether her speech therapist or the psychotherapist during college years. She could talk about work and her illness, big talk, but had little or no small talk. No boys asked her out, and there was no chance of going to the prom. "I was broken."

After school and college she found work as a secretary. Though she found less discrimination among adults than at school, she still missed out on many jobs because of both her looks and her voice. Eventually, she found a job and a husband—the first person who asked. It was flattering and she thought it would improve her self-esteem with a new identity as a wife. For the first time, in her twenties, she moved away from her parents.

Unfortunately, he was abusive, controlling, and coercive. Rather than improving her self-esteem, it was the reverse and, worse, she questioned her judgment again. She was married for a decade and endured it with alcohol, drugs, suicidal thoughts, and self-harming by cutting.

One bright part, apart from work—which she enjoyed—was riding her motorbike with a club, and it was a road accident on her bike that allowed her to leave her marriage. She fractured her wrist and needed surgery. Staying with her sister gave her the time to realize she could leave her husband. Alone once more, she had psychotherapy for four or five years. During this her therapist remarked that he did not always understand what she said. It was the first time anyone had said this for years, and she was transported back to her childhood shame.

She decided to go to university and read for an undergraduate degree in accountancy and then stayed for a PhD. After that she has taught in the community college for twelve years. This was online, and when her students asked her to record her lectures, she had to record and then edit them down Unable to avoid seeing herself, she was aware once more how different her speech was. Only recently in her late forties and early fifties has she accepted her speech, her cleft, and indeed herself:

> I denied having a cleft, that I was different, or even sounded different, until I was forty-seven. Though I knew I had a speech problem, I pretended I didn't and dismissed anyone who tried to tell me differently—even though my speech was a significant contributor to my lack of self-esteem, self-acceptance, and how I saw myself. I hated talking, so I reverted to writing.

Now a college professor, teaching face-to-face as well as online and making videos for students has helped her reduce her self-judgment about her speech, though she still struggles. She is about to do a local TED Talk on living with cleft and has been teaching herself to talk more slowly. With this she feels better about herself, which she now realizes makes people feel more relaxed and talk to her more.

As well as her accountancy work, she hopes to give back all she has learned from the cleft community, and has started to give her own therapeutic sessions online. She uses her own experience to try to build acceptance in others, to convince her clients that there is nothing wrong with them. One mantra is "No rulebook, no failure." She remains hard on herself, and though others tell her she appears strong, this is not how she views herself. She is determined that the next generation will not feel broken as she once did.

Marie

Marie works for a railway company, training beginners and planning new stations. Her last job, also in trains, included liaising with travelers on train platforms, which can involve explaining to irate commuters why their train was canceled. What is extraordinary is that as a child and adolescent, Marie could not speak before a group and for long periods hardly spoke at all. Born with a left cleft palate and lip, her mother left her in the hospital unwanted, so for the first three years she was brought up by her grand-mother. When she eventually was allowed back with her mother, if her mother's friends were around, they mocked her appearance and speech. The house was full of family photos, but none of her; she was "too ugly." "My mum did not want me. I was not perfect." Mockery soon turned to some-thing worse: both parents abused her sexually, physically, and emotionally. As a child she never smiled, never even learned to smile, and ended up with complex post-traumatic stress disorder. Her father was initially married to her mother's sister before living with her mother, and Marie soon acquired a half-brother and a full sister. Neither had a cleft and neither was abused. At one point a social worker appeared, and Marie asked for a different fam-ily, but nothing happened.

Her response was repeated, massive outbursts of anger and tantrums. Once she hit her neighbor with a metal pole and on another occasion man-aged to knock a friend unconscious—some feat when she was only four. Marie described herself as being a "child of rage." She grew up in a small village, but rather than making people more accepting it was the reverse. With little kindness from other parents or her own, she would keep quiet since quiet meant no attention, no abuse, and no bullying. She looked dif-ferent and sounded different, and in making sounds she was very nasal, had a lisp, and would stick her tongue out to speak. People even mocked that.

Marie had her first operation at six months and has had nineteen more: palatal repairs, implants, lip repair, operations on the nose twice, with bone grafts and gum grafts. Living with cleft means constant repairs and refine-ments rather than any single, life-changing operation. After some of the operations, say, for a bone graft when her jaw was broken and repositioned and the shape of her palate was altered, she had "to learn to speak again." Marie's anterior vocal tract has had innumerable changes both during child-hood and later, which she has had to learn to use anew for speaking, eating, and drinking. At one point her jaw was wired together for nine weeks, so it was liquid food only.

She had "lots" of speech and language therapists who encouraged her to speak more through her mouth and less through the nose. If she had a cold, this was particularly difficult. In those days there were few resources; she remembers a balloon up against her lips to try to feel vibration as she formed words. Another exercise was to blow cotton wool across a table through pursed lips. Air would escape through her nose due to her soft and hard palate not closing properly. To this day, she has a phobia of cotton wool and hates balloons.

At primary school her math teacher used to say she looked so sad. "No, I'm happy, I like school." Despite the other children singling her out, Marie loved lessons and learning. School was more normal than home. The teacher once asked her why she did not answer questions in class when she knew the answers, but inquired no further. With her parents mocking her speech and bullying her constantly, she desperately needed outside help. When none appeared, she did not speak. Was it her cleft, her parents, or just her? she would ask herself.

At secondary school, age twelve, Marie had a palatal repair. Before that, if she drank milk, it would come down her nose—her party trick. Despite all this, she was top of her year for math and English.

Because of her voice she was always conscious of whom she was talking with, knowing they would be judging her. She could talk with small groups but never large ones, always nervous and self-conscious. She'd rehearse the school play each year, loving it, but then never perform on the night. This happened each year from age seven to when she left at sixteen. She was good at math and often knew the answer but would never put up her arm to say it aloud. She would talk to her grandmother, to her one or two friends, and sometimes to her brothers and sisters, but to few others. Even at school reunions she would talk to her two friends but not to the other twenty-eight or so in her class.

At age nine Marie began drinking and later in her teens became anorexic. She left home at sixteen, staying with friends, leaving her parents "without a backward glance." Several overdoses and hospital admissions followed before she found a job, collecting glasses and cleaning in a working men's club. Two years later they persuaded her to start behind the bar serving drinks. In the club there was little etiquette or judgment—just working men wanting a drink. Gradually, she learned the rough old art of conversation and with it the confidence to answer back. Indeed, she became quite chatty during her bar work and would talk at 100 mph. To think where to put her tongue, she would daub Marmite on the roof of her mouth and touch it with her tongue each time. She was off and running.

Next, she took a job working night shifts. One night on the way home at 4 a.m., she was attacked and raped. Her attacker had six previous rapes and six robberies to his name. It was then that the police looked at her record and found reports of parental abuse and asked if she'd like to prosecute them. At that time, she had a hard enough time dealing with the rape, so said no. She developed a psychosis, was admitted to hospital under the Mental Health Act and admitted to a secure psychiatric unit, which precipitated anorexia once more. She was an inpatient for nine months, with therapy eight hours per day.

All this before she was nineteen. Three years later, Marie found she was pregnant and had her son. When he was six, she was pregnant again. Her husband left her soon after her daughter was born, not wanting much to do with them. Anyway, she preferred single parenthood to an argumentative relationship.

Over the next years in her twenties, she built a life. She worked for a medical laboratory service, initially filling in for people off sick and was so successful that she was promoted. She moved into mass transport and a number of jobs in train companies. Promotions again followed.

Now in her fifties, she can look back without too much pain. She caught up with her mum twenty years later and asked why did she not protect her more. All her mother could say was that she loved her husband too much to stop him. Subsequently, her mother drank herself to death. Marie's father was tried at the Old Bailey, imprisoned, and then released on the sex offenders register before he too died.

Marie still does not like people to look at her face. For one thing, if she has one operation, then it makes something else more obvious. Her last operation straightened her teeth and mouth, and now people realize her nose is not straight. Her voice does not bother her much now, even though people on the phone often ask for her mother. Once, when dealing with a suicide on the rails, even the police asked to talk with an adult. She still doesn't like new groups of people, especially larger ones, worrying about her looks and her voice. She also wears hearing aids since as a child she had glue ear (where the middle part of the ear fills up with fluid), which is common with cleft. Naturally, as a teenager she refused to wear them.

She often leaves them out at work too, allowing more focus on her work and the hundreds of emails she receives each day.[4] Now, "if I could change anything it would be the deafness and voice problems, not the cleft." The latter is an appearance and relatively fixed; the former needs effort and performance each day.

As her story tumbled out, I wondered whether I should tell it. I had, after all, approached her about the effects of cleft on her speech, and she had opened up to me about other matters after only a few meetings. She just shrugged: "Stuff happens." Indeed. Her experience—though appalling—is not unique. Many others with cleft reported bullying. Jessica, for example, emailed that she had been thinking about the speech aspects of her life and their "turmoil or sadness."

Jess had speech therapy from as early as first grade. "I remember being bullied, starting in the first grade. I walked home from school every day with kids making fun of me and my uncle would beat up the boys, but that never stopped the constant teasing." Her mother would make her repeat words she may have said wrong. But she still has some problems. "When I say 's' and 'p' or 'v' I still have to repeat my words, but I am sixty-one years old now so it is not a problem—unless the person makes fun of me with what I said."

Jess hated speech class because she was called out of the class and the kids knew where she was going: "I never told them, but they just knew." She stopped speech class in tenth grade, finally deciding she could take no more. After that, "I very rarely spoke out loud. I was so afraid of someone making fun of not only my speech but my mouth, so I was very quiet in school."

She would go home and cry herself to sleep, wishing she would die during the night so she did not have to return to the "war zone." Day after day, year after year, she endured the bullying, and not only from children: one elementary teacher was terrible. "She noticed I put my hand over my mouth to cover up the scars so no one would make fun of me and one day she yelled out to the entire class, 'Look at the baby sucking her thumb.' I was so humiliated, but I still covered my mouth."

In her early adult life, she became addicted to drugs and ended up going to both Alcoholics and Narcotics Anonymous. It took her a year to speak out at AA, and once that hurdle was past, she found people accepted her, flaws and all, without judgment. She began chairing meetings and became more comfortable speaking and telling her story. Now she can speak to a crowd of people without worry. Her problems have been as much to do with her appearance as her speech. But she does not mention problems with eating or drinking; her difficulties have all been social, the result of how others view her and their behavior toward her, not due to any of the physical inconveniences of the cleft.

Data

The amount and depths of prejudice toward those with speech difference is difficult to know, but for those with visible facial difference the figures are startling. A report by the UK charity Changing Faces, which supports those with a facial disfigurement, highlighted the results of a survey in the United Kingdom in 2017 of over 800 people. Of those with a facial disfigurement, 80 percent had experienced comments or unpleasantness from a stranger; half the schoolchildren with a disfigurement had experienced discrimination; and 80 percent did not apply for a job because of their appearance. More than half thought their condition hindered their career. Despite there being long-established links between disfigurement and emotional and psychological well-being, 75 percent had been denied medical or surgical treatment for their condition on the basis that it was "cosmetic" or unnecessary. Half had felt vulnerable on public transport, and 90 percent of those using dating websites had had uninvited, unpleasant remarks about their appearance from other users.[5] In a subsequent report, broadly similar results were found.[6] Sixty percent had experienced hostility from strangers, 36 percent job discrimination, and 30 percent felt depressed by their visible difference.

In the whole population of disabled people, who make up one in five of the population by legal definition, there are higher rates of domestic abuse, for longer periods of time. This abuse is more severe and frequent and in wider contexts and by greater numbers of significant others, including intimate partners, family members, personal care assistants, and healthcare professionals.[7] People with disabilities are often more vulnerable in their circumstances and in their ability to defend themselves, or to recognize, report, and escape abuse. Unsurprisingly, women are more likely to experience sexual assault, physical assault, and domestic abuse. One report from the Ann Craft Trust, a UK charity supporting the victims of these, estimates that in 2018, 16 percent of women with a long-term illness or disability had experienced domestic abuse, compared to 6.8 percent of nondisabled women.[8]

So much for appearance and physical impairment but for many with cleft, speech is just as much a concern, even though it is not always notice-able. Gareth Davies, who for many years led the UK Cleft Lip & Palate Association (CLAPA), before moving to head the European Cleft Organisation, once said to me,

> I don't think I am alone in stating that for me the most difficult part of having a cleft as a child was "sounding different" and all the repercussions as a consequence. Interestingly, I have found similar things adapting to life in France, and while my French is now pretty good the accent is still something that can cut me out of casual group conversations. I have in fact spoken in lectures about the psychosocial similarities of "cleft speech" and trying to get by in a foreign language.

He told me that at summer camps for those with cleft, the first thing he does is to send off the teenagers to put on a play. This is primarily so they realize that others sound like them; they already knew what others looked like.

When abuse or assault is involved, then there is legal protection, of course, though the reach and effectiveness of police and other services in such cases is far from perfect for many reasons. In the United Kingdom, prejudice against facial and other disfigurements was included in the Disability Discrimination Act. Changing Faces and its founder, James Partridge, was influential in this. Those with a disfigurement, or "an impairment," which

1. has a substantial adverse effect on your ability to carry out normal day-to-day activities; or

2. consists of a severe disfigurement. This applies even if your disfigurement has no impact on what activities you can do or on your mental health.

are covered, though the definition of a severe impairment depends on the nature, size, and prominence of the disfigurement and its effects on the person.[9] The Americans with Disabilities Act defines a person with a disability as someone who

1. has a physical or mental impairment that substantially limits one or more major life activities,

2. has a history or record of such an impairment (such as cancer that is in remission), or

3. is perceived by others as having such an impairment (such as a person who has scars from a severe burn).[10]

This may, however, leave open those cases where an individual may feel discriminated against despite little physical or aural abnormality being acknowledged by others. Of course, the law cannot prevent discrimination, though it can point toward a better direction for society. One hopes that the present and future generations with cleft and other differences will live easier lives than Jane and Marie. It was clear how successful Jane had been professionally. I asked if conversation was easier and more fulfilling: Did she chat, gossip, shoot the breeze? Her reply was brief as it was stark:

> My social life is nonexistent. I'm alone. I've not dated for ten years. I have no close friend, though I would love one.

She has occasional lunches with a few people, but they are not good friends.

> The possibility of rejection and its shame lives with me. I'd like a best friend; I'd like a relationship.

Was it the cleft or the speech? "Both." She cannot like her physical appearance or her nasal voice. I tried to suggest that, seeing her, appearance was hardly an issue. She agreed, but while her face was a constant presence and less intrusive, voice depended on speech and context and on performance and projection that she had never really mastered or felt confident with. Now in her fifties, Jane was still trying to learn these skills to converse with and engage others. A successful college professor, work was easy; what she lacked was the social, the casual, the simple (or complicated) acts of chatter and gossip. The rules for work interaction are laid down by the subject, the relationships and hierarchies measured in years and identity badges; these she had mastered. It was the less defined, more complex, and flexible interpersonal and social speech acts and games between friends that were so difficult.

Marie has established a life, brought up two children, and become much admired in her chosen field. But she's remarkable for one other reason too. She knew when twenty weeks pregnant that the daughter she carried had a cleft lip and palate—"I just knew"—and this was confirmed by a scan a few weeks later. Initially, Marie felt terrible; cleft can be inherited and she had done this to her precious child. Then she realized that if anyone knew how to manage it, she did. She took lots of photos of her daughter as she grew, before and after the inevitable operations. She also knew that even if her daughter had 101 operations, she would still have a cleft in appearance and in how she sounds. Cleft is her, but only a small part of her—she could show her daughter that. Her daughter was fortunate to have such a mother, expert, and mentor. Unlike Marie, she rehearsed and performed in the theater at school; unlike Marie, she ended up as senior head girl; and unlike Marie, she is now at university. Stuff happens, yes, but Marie has turned her terrible stuff into gold.

As with many other conditions, picture someone with cleft and it is likely to be a child. Implicit in this may be the expectation that with surgery the cleft and the visible signs of it will be diminished or even removed. However successful surgery is, though, the person is always likely to think of themselves as having cleft.

Sara has five degrees and speaks four languages. A professor of liberal arts, she has lived in the United States, Costa Rica, and China, and regularly visits Austria, Italy, and England, teaching students how to be, like her, "a citizen of the world." For much of her life she has also pretended—"de-associated"—that she does not have a cleft lip and palate. She did not want to be a victim, falling into the passive role that implies, having experienced something of that during some of her fifteen surgeries (so far). She remembered how her self-esteem had been damaged by her look and sound, mainly because of others' responses. People would repeat what she had said to check they had understood, and whether family or friends, each time it reopened the wound. It was only in her thirties that she began to feel part of the cleft-affected community and to feel ready to explore her condition, and what she had learned along the way.

She realizes now that she will always be cleft-affected. Despite improvements in medical and surgical practice, she will always remain unrepaired in some respect:

We cannot be fixed. I still cannot blow a balloon up. My education, my relations with my daughter, my divorce: all roads lead back to cleft. I have had therapy but, like the roof of my mouth, there is still an unrepaired wound. Now I want to explore and reframe my position, and if I can pass on my knowledge, not as victim, not as survivor, but simply as a human being, navigating my unrecovery.

At its core Möbius syndrome involves the congenital absence of facial expression and ability to move the eyes laterally (due to the malformation of two cranial nerve nuclei, the sixth and seventh). From birth, people with Möbius cannot shut their eyes, shut their mouths, smile, or move their faces, nor look side to side. Some also have problems with their hand and foot development, with their teeth and jaws, and with tongue movement and swallowing. They can also lack spatial awareness. With such a wide variety of bodily problems it is unsurprising that they can also have learning difficulties, though with additional help these can be overcome. Social development can also be problematic, since without facial expression their intersocial skills may lag behind others.[1] Without lip movement, it also is not surprising that many of the consonants are difficult to pronounce. Though their experiences go way beyond this, people with Möbius can reflect on their experience of speech without being able to move their lips and much more.[2]

Jo is now in their forties, and well placed to look back on their childhood and on their acquisition of speech.[3] One of the first problems was saying consonants without moving their face and lips. "If you go through the alphabet, about half the sounds use something difficult to say, 'm' and 'b's etc., which is unfortunate given the name of the condition." Like many others, they developed speech late. Even so, they have no memory of how they did it, nor how they make the sounds now.

> Unfortunately, this is where there is a big hole. I have no recollection of my speech therapy. I can tell you the name of who checked my eyes, who did my foot operations when I was three, all my various checkups. It had to be a game with all the different appointments—that was my life—but nothing on speech therapy— nothing. I was young, but it was after I had developed memories for other events.

My questions led them to wonder why their memory for this is so much worse than for other matters, even unpleasant ones like operations, but it remains unclear.

Now, as an adult Jo's speech is extraordinarily beautifully articulated. Jo has had a variety of jobs and has given lectures on facial disfigurement to speech and language therapists.

> They just look at me, wondering how I produce my sound. I don't know. All I do know, because I like language and foreign languages, is that I find a way to imitate it. I started German recently, and said to the teacher, "I will not produce the sound in a conventional way, I'll have to hear it and then work out how my mouth can replicate the sound." I must have done that. But I don't know how I did that then nor now. I do remember my diction was less clear when I was younger, and I would be asked to repeat.

Listening to Jo, the Bs and Ms tumble out apparently as normal, or as near normal as we can detect as speech proceeds. One place to look at how Jo might form some consonants is in ventriloquists' instruction manuals. These suggest a variety of ways to produce labial sounds—the troublemakers, B, F, M, P, V, and W—without moving the lips. In addition to these, people with Möbius can have tongue problems, so those sounds that involved the tongue on the roof of the mouth—D, T, N, S, and Z—can also be a trial.

The manuals suggest a variety of methods to give the illusion of labial sounds. One is substitution of one sound by another that is easier to produce; instead of the sound "W" in a word it is suggested you make a "Huh—oo" as one sound that flows immediately into the vowel that follows. Instead of B, saying a G or D is suggested, hence the usual phrase related to bad ventriloquists, "Gread and Gutter." Another substitution is to replace B with a "geh" (soft Spanish g) sound at the back of the throat, So "banjo" would be "gehn-jo" in sound; M becomes "nah" or "neh," so "master" is "nah-ster." Ventriloquists are helped by our own expectations of language and words, so we fill in the sound we hear with that we expect to hear.[4] Jo agrees: "People know what they are looking for, and so hear M, N, D, L, B." but there is one other method suggested by ventriloquists. "When substituting 'B' with a 'D,' notice how your tongue hits the roof of your mouth. Begin to change how the tongue comes into contact with the top of your mouth. Instead of where it usually hits, move it forward so it contacts the ridge right behind your teeth. The sound changes. Keep playing; it takes time, but you will eventually get a 'B' sound." Alterations in the way the tongue contacts the roof of the mouth can modify the sound. However it is done, many with Möbius have such excellent diction you'd be hard pressed to spot their impairment over the phone.

As a child Jo read widely, which helped them develop a fine vocabulary and good diction. But words took them only so far:

> At five years of age the only talking I could do was big, about operations, say, to doctors; I could only talk to adults, about my bits not about me. I could also talk about books. Adults were my friends, not children. I just could not do playing with the other kids. With adults I would have a conversation, but with children I was a bystander. Children had another language, a word language, a body language, a facial language. They run around and jump up and down and I could not do that because my legs did not work and because of my lack of balance.

Jo had big talk but no small talk. With their poor balance and coordination, games were out, so there was no rough-and-tumble childhood. With their eye gaze problems too, Jo could not judge distances or moving objects, so could never catch a ball. They also needed many operations and medical treatments:

> I did not do ballet, horse riding, etc.; I did hospitals and operations. I had the eye doctor and the foot doctor and a speech therapist—who I don't remember—and a face doctor.

Jo's myriad of conditions, involving feet, mouth, eyes, hands, and coordination, and the endless round of visits to doctors and therapists led Jo to an unusual way of viewing themselves:

> I never thought I was a person; I used to think I was a collection of bits. I thought I had all these different doctors to look after all the different bits. At half term other children would go off camping or swimming courses; I would see the doctors, this one, then that one. "Jo" was not there; that was a name people called the collection of bits.
>
> I liked my spirit because I was strong as a child; I liked my brain; I loved reading and read very early on. I could think and dream and imagine. I always knew there was something strong inside that I had a mental dialogue with but it was not the physical body; it was very separate from the physical.

This disconnection between their whole thinking interior self and their failing somatic bodily personhood led to other extraordinary consequences. They lived in a correct English family not given to communicating their inner lives:

I am a word person and I think in words and did so even before I could communicate or articulate. But then I did not have someone good enough, receptive enough, to speak to. I was in this grown-up head so it often would have not made sense to say it out loud. An early memory: about six at my brother's christening [in the] family photo they put me in the front, being a little girl. The sun was in my eyes, which is not good with Möbius and I just thought, "Why are they all so stupid that they cannot even understand that?" There was no point saying it; I just thought you don't realize. I totally knew my own mind.

A year later Jo needed a foot operation:

When I was seven I stopped walking because my feet were so bad and I had to go to school in a wheelchair. After some surgery I could walk but I never told anyone I was in pain. People don't ask little children. No one asked.

It was as though Jo could use words and speech for external events and for their precious novels and books but did not realize speech could express an inner life as well. This Jo kept to themself, living—as another person with Möbius said—"entirely in their head." In tandem with this, something else was missing:

I did not express emotion. I am not sure that I felt emotion as a defined concept. At my birthday parties I did not get excited. There were people around excited but I followed what they did. I don't think I was happy or even had the concept of happiness as a child. I was saddened by being in pain or having horrid things like a blood test. Sometimes I would cry but even that would almost be a delayed reaction.

Fortunately, as we will see, Jo has developed a rich emotional life subsequently. Somewhere in the somatic problems of the condition, whether the absence of facial expression or gesture or vocal prosody, the reduced ability to express or communicate emotion appeared to be coupled with an absence of or reduction in emotion as an experience:

At the time I just endured it. I could not express my feelings. Sure, I saw children in the playground laughing but I always related it back to the physical. I did not know about an emotional world. I thought it might be related to my legs.

I knew that being happy was something I couldn't do. Everyone told me I couldn't smile. I never got excited at Christmas; I watched others being excited. I verbalized it but not in an emotional way. I knew things were not as they should have been even though I did know how they could have been different. I did not realize I was unhappy; I just was.

At their first school, a girl said, "Oh, you can't close your mouth." Jo thought, "*That's* what it is. It was the missing link. I knew my mouth was different because I saw it in photos, but no one had said it in so many words. It was interesting." Once, a mother at a birthday party told Jo to close their mouth when eating. Jo did not know what to say since they knew the mother would not understand that Jo never closed their mouth. So they cried so they could leave the party. One girl at school had diabetes, so they became friends, fellow outsiders. Later at secondary school, another friend lived with cerebral palsy and they compared health things, though Jo realized—even then—that this was a narrow base for companionship. Another friend just chatted. Chatting, social back-and-forth conversation, was not something Jo did. "I could see what she was doing but I did not know how to do it, and so had less to say."

While at school, for the first time Jo began to realize that other people's faces were expressing something inside:

> This made things easier since you would know how they were feeling. I was conscious of people not knowing about me, what I was thinking and feeling. A person in my form was fairly moody. The teacher said it was because she has red hair. I was aware that people did not know with me but then I did not know either; I did not have a clue. I thought the only way I could show people was through speech. Nonverbal clues were divorced from me. I did not know they conveyed a message. The more I could see and read other people's faces, the more they were different to mine. I did not like my face and I did not think other people liked my face. I never stared in the mirror because I did not like what I saw. I did not want other people to look. I did not want to be noticed. I did not know what facial expression did and was for. I did not understand it showed feeling and expression.

Foolishly, I asked if Jo had tried gesture.

> I did not have gesture because I had not learned it. When I was a child, I could not gesture because I was a collection of bits. My body was not me, so expression in it, with it, through it, would not be from me either. It was not a joined-up feeling. There was huge bit missing; with the lack of balance, mobility, and problems with coordination, you don't get a sense of self. . . . I could see everything and wanted to communicate but I could not do anything. It makes you so different. The adults may have been trying hard but I could not give back.

Though Jo's lack of social skills and playground movement prevented much of what school is about, they liked lessons. Learning fed Jo, Jo the bright disembodied mind, like nothing else. Now, fortunately, Jo has a wonderfully rich emotional life and experience. Its development may have begun in their adolescence. Deeply unhappy, they would talk with a relative, Catherine. One day Catherine listened to their troubles and said nothing; she just gave Jo a big hug:

> No one had given me a hug before. I was so shocked to know what physical warmth was, to be approached through the body, not through the intellect. This embodied experience was extraordinary, and the next day it startled me out of my wits. There is a whole new something out there I knew nothing about. She loved me and it was not conditional. It meant there was a worth and value to my life; at that stage I could not understand why anyone might love me since I did not even like myself.

However, this did not help with Jo's contemporaries. Jo kept quiet, studied hard, and looked forward to university. But when the time came to apply, no offers came in; the school's report had been "unhelpful." Though Jo felt badly let down, they knew that they could reapply after their A levels rather than before, which was more usual. Jo spent three days phoning thirty to forty universities and pleading, admitting to some health issues but saying they were prepared to work hard. Two accepted them, and they made their choice:

> I was aware it would be hard away from home. I could picture myself going to lectures and doing the work but I was not sure how I would make friends. I knew people went to university and had a good time, went to parties and got drunk, but I was not sure how I was going to do that.

Her mother left Jo in their room on campus the night of the Freshers' dinner. Jo had never been anywhere new alone. They left the room and found two girls coming downstairs, so asked to go with them to the meal. As the two talked nonstop to each other, Jo just watched and listened, having no chit-chat themself. The girls did not tell them to go away, so Jo went with them, watching and learning. After the meal they suggested a drink. Jo had only ever been out with their brothers. At the students' bar they fell in with a group, and though not really Jo's sort they allowed Jo to stay. Jo soon realized that if they asked the right questions, others would chat and this created a dialogue. Initially, Jo found it difficult to immerse themself in a more social, gossipy language; small talk seemed superficial but they soon found the rewards worth it. For the first time in Jo's life,

> There was no feeling that people talked to me because they had to. There was no animosity; I did not feel judged, I was less a leper. Some people wanted to spend time with me, to go shopping, go out in the evening, sit next to me. No one had ever done that before.

In one lecture series Jo and a friend used to sit at the back and whisper and giggle in a way Jo had never done before. Suddenly, it seemed, Jo knew lots of people unconcerned with their face. Jo would listen to a radio pop station to hear what was hip so they could talk about it with others. Having gone to university to read for a degree, Jo found they were actually studying people. "Everyone came to me to tell me things. I was like a priest. I landed up with loads of stuff about others and about life." They learned much about many different experiences and interests, for instance, about abortions and the trials of single-parent families. For the first time, Jo also had a choice of friends. They bought hippie clothes like everyone else and started to develop a character:

> It was maybe artificial but I could design my own. Most people's evolve as they grow up; mine I picked. I wanted to be someone whom people would like—before people had not liked me—sweet, gentle, and likable, inoffensive, reliable, so that is what I became.

Above all, Jo wanted not to stand out. They learned body language and gesture for the first time, and developed a broad circle of friends where they could talk about different things with different people. They got a job in a café where the "cool" people were:

> It was no longer me on the edge not knowing how to get in nor what to say; I'd be in the middle giggling, being teased, being normal. One tall, gorgeous guy who all the girls adored and who was a total rogue was so sweet to me. He would pick me up and throw me around and hug me. It was a total flip from my previous life.

It was during a summer job that Jo met a doctor who was able to explain, for the first time, what Möbius was. Despite being in their early twenties, this was revelatory for Jo:

> He drew pictures of what was not working and showed me that it was the face and facial expression. I realized it was not simply that my face was a different shape; it was what it was not doing. The things that had been focused on as a child, the eating and the speech, had been sort of perched there without focusing on the face and its expression as a whole. I had no idea that people give off nonverbal signals.

During their time at university Jo learned how to dress, how to chat with people, how to laugh (Möbius style, by making a "hee-hee" sound and shrugging the shoulders up and down), and how to gesture. They invented a whole new Jo, more social and independent. But while this allowed them to fit in, it was skillful imitation, for they did not feel what they expressed.

The three years at university rushed past. Back home with their parents, it was difficult to be the new confident Jo when the family was so unused to it. Jo applied for jobs, application after application, without an interview. Though not keen since it had a bad press at the time, Jo decided to try teaching English as a second language; after eight months at home they needed to escape. Jo knew that they interviewed well, but several language schools turned them down. One even said that even though Jo was very good, they were not going to take them because of their face.

Eventually, a school agreed to take them, and after that Jo found a job in Portugal. When Jo got there the job disappeared, only to reappear when they returned to England. Undeterred, they tried Romania; Jo had always loved their students: they had more fun and more parties. Jo found a job teaching students and boat-builders and soon picked up enough Romanian. Their face was noted but not a problem.

It was there that Jo's education went way beyond Romanian and language itself. At university Jo had learned a lot of imitating, mirroring, and copying, but had not felt what was communicated. When speaking Romanian everyone was so dramatic. "If something awful happens, then the world is coming to an end, and if it's fine, you party all night." Its culture allowed Jo to develop in ways they never knew existed.

Because of the cultural "upregulation" of feeling into gesture, *I learned to feel*. I am not sure how I mapped gesture and feeling onto my body, but I was starting to feel then. I could feel really ecstatic, happy, for the first time ever. Before without the expression I found feeling difficult. Once in Romania I certainly had the means, the channel and the vehicle, and the feeling. Previously, my thought was frigid or cold. I needed the continuation of a thought into real time expression within the body.

I was an intellectual at university. In Romania I experienced emotion. As a child I used to play a musical instrument with emotional expression, but the emotion did not really come from within. Once I could express there was no stopping me.

Now I could talk about how I was as a person. Instead of facial expressions, I used my hands and shoulders, and my voice, both in its tone and what I say. I was twenty-three and had just started to learn to do that. It was like learning another language but without a dictionary; I had to invent it and work out what worked. I had to learn register, the use of the appropriate language and expression for the situation and for each audience.

I learned Romanian in two months but—more—they are so theatrical in their emotional expression. In Romania, ironically, though they use gesture a lot, they talk in a much more monosyllabic way than here.[5] They are not as musical in speech. My voice is melodious and I had begun to use it to control people's response. So in Romania they all talked about my voice and loved it. That was a Eureka moment. My voice could be my thing, my tool, my vehicle of expression. The voice was the link from me to other people for feeling and for emotion. The language and words don't do it. I had those at university. It was the voice, the melody going with the embodied gesture that completed the circle. As I experienced emotion more and expressed it more, then the language—my language—my words and their expression changed as well.

I do this thing and they know what I am feeling . . . experiencing emotion, amplified from them. Like iron filings to a magnet, I was drawn to it. I was in a totally different context; I could be me as I wanted and it seemed too irrelevant how I looked. I used to phone my mum each week and she said, "You're different."

This is not to say that with emotional experience, Jo's expression was the same as others. Even now in their forties, all Jo's gesture and prosody is performed consciously:

Everything I do, I think about. . . . All the things I am doing, whether turning my head or moving my hands, I think about. I still construct it all very carefully. I am very aware the whole time. I am looking at you to make sure I am engaged with you. My voice I use to reinforce emotion. When I want to make a point, I really concentrate on my voice and intonation. I have to monitor all these things constantly the whole time. None of this is automatic. Because I have to think about every sentence, and how I am to deliver it, my verbal fluency probably slows down. It is tiring but it is what I have to do.

As we chatted, I mentioned Robin Dunbar's theory of language as social glue and how a speaker could reach a larger number of people than, say, by grooming. Jo was aghast. Jo tailored their speech patterns, volume, prosody, and gesture *individually* for each person or small group of people they met. Jo disagreed with the idea of grooming lots of people in the same manner. Merleau-Ponty wrote, "I exist in the facial expression of the other, as they do in mine."[6] Jo is aware of how important the facial responsiveness of the other is, and they not only tailor their responses but continually monitor how successful they are and adapt accordingly. Jo is now aware of how their face reveals little or nothing. Instead, Jo exists in their speech and in language, in their vocal range and in their gesture, and these are tailored continually to their audience, either a person or more generally a larger group.

Even now to consciously think about speech and prosody, and about gesture, and to be so aware of how people are receiving these has several consequences for Jo:

I think this makes me very vulnerable to other people's states of mind, whether angry, sad, or happy. I pick up signals and it can be very unsettling for me. I have to remind myself that it is their stuff, not my stuff. I have found that very hard and only recently have I found a way which works for me. Otherwise I can be caught up in others' situation.

With Möbius you have to be much wordier and articulate, and this can be hard and tiring. For me the word is stronger than facial expression. When I teach language we talk about the nonverbal, but what I am interested in is the non-facial—gesture and especially tone of voice. Without these thought is impoverished, as is language without gesture and voice, and thought without language.

As a child I did not know I was missing out. To reach out to others I needed embodied expression—feeling. As you grow up the social feedback from others has far more meaning than as a child. A meaningful smile from you triggers an emotional response in me. As a teenager I was articulate, but this was insufficient. I had words, I read, I was eloquent, I could talk to adults [and] they could understand me, but none of this helped.

Sitting with Jo, richly emotional and acutely aware of their inner feelings as well as keeping a trained eye on yours, is stimulating and enervating. By constantly monitoring their success with others, it is as though Jo is aware of their own feelings in a heightened and enriching way. There is less reserve and, rather, a life-affirming presence without facial expression but with gesture, prosody, and phrasing of the voice, and with a keen intelligence exploring and relishing their feelings and yours:

When I was playing emotional catch-up, I never associated that with Möbius. Now I am happy most of the time. An emotion is what you feel, a state of mind, what you think. If it's a nice day you can think that's a nice day, warm blue sky, green grass, nice view, and you can think happy . . . but to be happy—an emotional experience—you have to feel it. Sometimes I feel happy and I smile in my face but you don't see it.

Now, when I express being happy that has to be vocal and intellectual. . . . There is an element of artificiality of the expression but not the feeling. Even with emotion now I have two sorts of happy: intellectual happy to express, and happy happy. I can be happy, not think happy, but to express I have to think. As a child I did not know I was missing out. To reach out to others I needed embodied expression—feeling.

Jo is still getting there. Recently, a good friend commented on a family relationship and the lack of emotional vocabulary in another friend. Jo's first reaction was not a word but an emotion, to feel sad and exposed. "For many years, I would have had to have a word response first but now it was an emotional one; my first reaction now was to cry. I was pleased afterward.

What came first was the emotion—embodied emotion—and not as before the words. That's really important."

The effort involved in expressing themself continually to others and in turn to be so aware of others' reactions is considerable:

> It is very difficult to describe the energy I need. It is exhausting, even just talking to you now. As soon as I start exteriorizing myself it is an effort, because I have to think about gesture and about what I say and how I say it. If I don't know the end of a sentence, then I slow down. It's not natural with Möbius.
>
> I do the full-on much of the time at work and in front of people, always full-on. Still after all these years, every time I meet someone I am managing how they perceive me, how I can try to help them, how I can help them if they don't know me, how they might be shown how to respond. Any person with a disfigurement has to manage each social interaction. I have to do all this without the other people realizing each day, all day, every day. In spite of having a heavy bag to carry about, you can still live and enjoy. It is frustrating when people don't know just how heavy it is.

After their years of isolation, Jo is now able to see that Möbius has given them some things as well:

> I have an ability to see things and value them more. I don't get fussed by little things. My piano this week would not fit through the door. So? Having not had friends as a child, having so many operations, something now has to be really bad for me to be worried about it. I only get worked up by the big things. Möbius has taught me to value the simple things that make you happy. Don't wait for dreams to come true; value the nice book, or the walk, or a shared meal. That makes me more grounded and stable, and I think I get more pleasure on a daily basis. In fact, I think I am happier each day than my friends because I have known the pain of isolation. After fifteen years of life without having a giggle or a conversation or not going to cinema because you have no friends, then every day is a pleasure.

Picture Jo saying these words, not as a recitation but in conversation, with changes in tone, volume, cadence, and prosody, and with emotion and humor. Try to imagine that all these were learned and are still requiring attention. Just as the losses of Möbius can be difficult to comprehend, so are some of Jo's daily compensations and workarounds, as a child and now four decades later.

8 Just like My Elbows: Parkinson's

Parkinson's is a progressive condition with a classic triad of symptoms: shaking, stiffness, and difficulty with initiation of action. Its grip on a person, though, is often far more pervasive and the voice does not escape. What is curious is that while some people are aware of this effect, many others are not. One woman with early Parkinson's who realized her vocal problems described how it made her sound:

> Somnolent and dull. I have heard myself on recordings leaving messages, for example, and I sound zombie-like with big pauses before words, sounding unnaturally slow and effortful. I struggle to say certain sounds, like D's and T's. When I anticipate them coming up in a sentence, I get anxious and stutter slightly. For example, just saying a day of the week I will often pause, mid-word, before I can say, "Tues . . . um . . . Day" or "Wednes . . . d-d-Day."
>
> My lips have begun to remain closed and stiff when I am trying to say something. My voice is lower than it was. I sing an octave below my old voice—so I sing with the tenors rather than the altos!

She knows what she wants to say but has to concentrate so hard on the mechanics of articulation that meaning may be compromised by the sheer labor of speech production. Tone, gesture, and facial mobility are suppressed, and she can often feel that she has failed to communicate. "It's like having a complete picture in my head but being unable to get it down on paper. So frustrating." At one dinner party, the hostess, seeing that Mary wanted to contribute to the topic, hushed the others to make space for her. But once her simple remark had been made, it sounded lame and pointless, out of kilter and awkward. Timing is all.

Sometimes she loses the thread, which she finds unsettling. Nagging away are thoughts: "Is my mind shrinking because I find talking so difficult? It's hard to think in a complex or abstract way when you can't actually say it."

Despite her difficulties with typing, she still writes coherent, funny, warm messages. She launches out on written sentences confident she can finish them, in a way she does not with the spoken word. Mary fears that she is losing aspects of language and wonders, "Am I developing those entire spongy, 'empty' zones one sees in images of the brains of Alzheimer's patients?"

Another woman, Lizzie, with more advanced and severe Parkinson's was able to reflect on years trying to outwit it. "You know how some children have absorbing, all-consuming hobbies? Well, I hated ballet, was scared of horses and couldn't sing (still can't). So, for a long while, I did not find my niche in the world." Instead, Lizzie was an early fluent reader. Her parents ran a Sunday school, so she quickly became understudy-in-chief, standing in when someone else didn't turn up. Reading including sight reading, something she could do and enjoyed. "I could read out loud confidently, so I was rather precocious in, for example, reciting funny poems at church concerts." In her teens, her mum saw a drama teacher advertising in the local paper. Lizzie went along and loved it. "I got on like a house on fire with the teacher and just loved this new hobby, and it was all-consuming for a few of my teenage years."

Her mid- to late teens were filled with learning poems by John Betjeman or comic speeches from Shakespeare, with weekly lessons in what used to be called elocution. "Rather than treading the boards in musical theater, it was just me and my wonderful teacher working out the emphases and the dramatic pauses in my pieces."

Next were competitions in public speaking, soliloquys from Shakespeare, monologues, and poetry. She loved the rhythms and rhymes of Betjeman, Ogden Nash, and Joyce Grenfell, popular English poets and performers. Looking back, she sees this as a mixture of elocution, drama, and literature. Though she was not interested in going on the stage, she did enjoy performance, whether party pieces at church concerts or exams in poetry speaking or public speaking. Above all, she enjoyed the elision of good pronunciation with the elegance of literature, words beautifully written to be spoken and inhabited. She was bright and good at school, but her love was in performative speech.

Next were public speaking exams and competitions. She entered various high-sounding exams: the Poetry Society Gold Medal for Verse Speaking, and the London Academy of Music & Dramatic Art Gold Medals for the Speaking of Verse and Prose. In some of these you are given a topic on the day and sight unseen, with a few minutes to prepare a speech before delivering it to an examiner. It was not competitive but was satisfying to her. Public speaking was her forte and underpinned her self-image. Even when doing drama she preferred rehearsal to performance, process to product. When she tried amateur dramatics, she never wanted the performance to come round, since then the rehearsals would stop:

> I liked the rehearsals; I was never interested in actually going on the stage. We spent the lesson times deciding how to deliver a sonnet or whatever, and that was what gave me the feeling of something achieved that other teenagers got dancing in a tutu before an audience. I did do the exams and even those public speaking competitions, but a medal or certificate wasn't my main reason for doing them. I enjoyed having a copy of a poem covered in squiggles to remind me what to emphasize.
>
> I was super confident of my voice doing everything that I needed it to. So much so that I didn't need to think about it ever, at all. It was just there like my elbows or my spleen, and you never need to check with yourself if your spleen is working properly or if your elbows are going to let you down; they are just there, doing their elbowy stuff. My speech was more than automatic; I could trust it in every situation and I did not need to think about it at all.

She did well at grammar school and went to an Oxbridge college to read mathematics, which had always come easily to her. Though her college had mostly public-school students, she held her own at interviews or at high table with intimidating dons, in large part because of confidence in her speech. "I never doubted myself as a speaker, conversationalist, or teacher. Conversation came easily without any preplanning, and that helped in those awkward sherry party moments."

Despite a double first in math, she was self-aware enough to know that her math understanding ran out halfway through her third year; after that, she was learning the theorems but not really understanding them. She planned to get married the year after she graduated to Iain, a teacher. At that time in the mid-1970s, it seemed reasonable to live at home, do her teachers' certificate, and make her wedding dress. She'd be livid now if one of her students with such an academic record did that. Iain became a minister of religion, and they have spent thirty years in various manses, and now live, after their retirement, in the depths of the country. They have two children and three young grandsons.

First, though, were jobs and job interviews. Her teenage years delivering poetry and prose stood her in good stead:

I didn't have to plan or think it through in advance; I just opened my mouth and out came the right sort of words all clearly put across at the right volume. These ranged from tutorials or sherry parties with intimidating dons, to not needing voice projection advice after I had been watched once on teaching practice, to chatting with older church people with hearing loss and a soft spot for Iain!

Applying for her first teaching job, she had three interviews in two days. "The posh independent school and the fairly rough comprehensive both said to me, 'Why is a person like you applying to a school like this?'" So, she ended up at the grammar school. This was in-between on the social scale, and incidentally the same type of school she had attended.

She worked as a teacher of math all her life, ending up head of math in a mixed comprehensive with a large sixth form. She loved teaching, with its performance element, out in front of her students and aware that her success depended on how she put it over as much as the material itself. There she was comfortable, at home, herself. So comfortable that a sixth-form skit had her as Dave Allen, a much-loved Irish comedian of the time and famous for being laid back. She also loves people and chatting. "I just enjoy talking; all those incidental little conversations in the dentist's waiting room, the cinema queue, the pick and mix aisle in Tesco's."[1] She wonders if she values small talk more because her elder brother lives with high-functioning autism and when growing up he had none.

Her Parkinson's was diagnosed in her fifties after several productive years as head of department, though it probably began some time before. She developed a slight tremor, which she and her GP ignored. Then after a foot operation for a bunion, she had a strange feeling while walking, as though wading through treacle. She did not realize these could be connected, so she told different doctors about them. She saw a neurologist about a year later, by which time it was clear from her gait as she walked in that she had Parkinson's.[2] "I did not deliberately ignore these symptoms, I just did not make the connection."

Tremor and altered walking were her visible symptoms. Doctors, family, friends, colleagues, and crucially Lizzie herself all focused on the motor symptoms early on and finding the best drugs. Wrapped up in finding the medication that helped the major motor effects, she did not think about voice for quite a while, even though with hindsight she realizes it was her first symptom.

"No one, not even me, gave a thought to speech difficulties; the signs were there, but no one recognized them." One of the first signals of Parkinson's was in the classroom where her "teacher voice," above even her "social voice," in retrospect showed the first sign, albeit unrecognized. As a head of department she should be able to quell a corridor or quiet a room without difficulties, with either a look or a mild rebuke. But now she found her voice had changed, and she could no longer silence her students nor find the killer stare:

> In a classroom an experienced teacher is constantly performing and multitasking. Part of her armory is "the look," the silent warning to the known offender as she continues with the lesson. I lost that. Teachers generally keep that forever—even thirty years after he last taught Iain can summon up "the look" to frighten fidgety bridesmaids while he calmly continues with the legal wording of a marriage.

The loss bewildered her for quite a while.

> I had some very odd moments in front of classes feeling like a newly qualified teacher and not knowing why. Once a teacher, always a teacher, they say, often because of the glare or the icy tone of the warning voice. Mine seemed to have gone.[3]

She was very open with both colleagues and pupils, who were accepting and understanding:

> The loss of the teacher's glare and loss of voice volume meant I held the floor much less well in a classroom. It seemed fine with the sixth form since I did not have to do the discipline thing as well as the exposition or whatever. Teaching lower school classes seemed to be like fifty-five minutes of "don't ever forget, you're a person with Parkinson's" (PWP), whereas teaching sixth formers was "still me."[4]
>
> I told colleagues quite quickly about my diagnosis, but one of my first quandaries was the question of whether to tell the students. Someone wisely said one criterion to use was whether it made it less stressful for me for students to know. I did tell them and it does reduce the stress and seems very natural, and I even get interviewed for their health and social care coursework now: "living with a long-term condition." The only problem is telling subsequent years, as I expected them to know (the way secondary students know most things before the staff do) and in fact they didn't. Which is better as it's not news; it's just a fact they tuck away.

Some years later she was sent to occupational health since her voice was becoming softer and more difficult to decipher. They suggested she have a stand mike complete with four-foot-high amplifier. Since an inner-city high school has limited space and few empty classrooms, part-time teachers would trudge from room to room, glad when the timetable was kind and they could find a classroom to sit in rather than a cupboard. The corridors were also crowded between lessons, and she would have struggled with exercise books for marking, her stick, lesson plan, and a four-foot-high blaster; it was hilariously impractical. Instead, they settled on a small mike and amplifier on a lanyard around her neck, which helped for several years.

By then she was trying different drugs to reduce the worst of her symptoms—tremor and walking difficulties—to manageable levels of discomfort.[5] But it was the vocal effects of Parkinson's that she began to realize were going to be the problem. What had previously just happened without thought or attention—using her teacher's voice with its command and authority—had not only deserted her but also at times deceived her too:

> This is the nub of the matter. Partly I thought I was doing it right, loudly and with expression, and why on earth aren't they responding as they should? After a lifetime of teaching I knew the fault lay within me. I knew when I was more distant from a maths class, though I didn't think of it as distant; I thought I was in a parallel world that didn't intersect with theirs much. My speech did get very quiet and sometimes—this is tricky to explain—I did sort of talk *into* myself, and then I could be standing in front of a class and explaining something with little response, and I would be unsure if I had spoken it aloud or not.

She realized she would have to take early retirement. Gradually, over the next six years or so, she relinquished teaching the younger age groups, became increasingly part-time, and handed over the reins of her department to a new head of department. Parkinson's was stripping her of authority, assertiveness, and her professional self. She felt a

> loss of a major talent which defined me (to me at least) . . . loss of confidence, loss of assertiveness. Self-aware again, I knew I wasn't teaching as well as I had been able to. I was definitely past my best, as some scurrilous student put on ratemyteacher.com once!

With the school's help she cut her timetable down and down, and then took early retirement. "I sort of melted away. It wasn't at all hard to leave the school environment—by gradually shrinking my timetable I wasn't involved in much." It was very important to her to continue to be able to teach and contribute to the department. She did not think too much about the significance of retiring, and of retiring in the quiet way that she did; she was just grateful to the senior management for their pastoral care. She thinks far more now about the difference between her almost invisible slipping away and the razzmatazz there would have been if she'd really been able to put her stamp on the department.[6]

Retirement removed the stress and overwhelming fatigue of work, traveling, and managing her symptoms. "Fatigue: teaching is tiring, and keeping the lid on the worst of my parky symptoms all took it out of me." She focused on family and friends—her church gave her community—and on day-to-day living with Parkinson's and its myriad of symptoms. Not having to use her voice to teach was a blessing; she joined various groups in the village and was able to use her purely social voice more. "But it surprised me that social chat was as tricky as teaching for me; I hadn't had time to realize that when I was still working."

She always felt she had three overlapping strings to her bow—performing, teaching, and social conversation. Retirement wiped out the middle one, so her social conversation became much more significant. "In my little world there were no exam boards to meet with; they were replaced by the village Knit and Natter group." Though she had talked lots at work, retirement showed her just how complex casual conversation is. At work her talk was of timetables, student progress, and marking, and more with formal, even if unwritten, rules of communication. In contrast, "nattering" about the messiness of people's lives was a freer form. Whether choosing the right moment to make a joke, coping with the unpredictable timings between individuals or with people interrupting others, nattering had fewer rules she could cling to.

"I really don't think that deeply. I just got on with being retired, and as my last few years of teaching were part-time, I had joined some social groups in anticipation of retiring. Back when I retired, I didn't look too parky, and I didn't sound too parky. ('Parky' is my shorthand.) But now nearly a decade later, I look the part and at times I sound like a Person with Parkinson's. I think it will be harder to join a new social group: my disabilities precede me."

Just as some people assist others with their mobility problems inappropriately, so she found that friends and family were likely to jump in and speak for her because they knew her better and could anticipate what she was trying to say. They would almost automatically explain what she was driving at, which she found at times "rather irritating." Ever fair, though, she wonders if she was maligning her nearest and dearest. "I found it hard to be gracious if the punch line of an anecdote was snatched from me."

It felt worse because they were only trying to help, but she found such interruptions were the worst thing for making her lose her thread. What was previously automatic—thought into words, listening and replying to others—was becoming more difficult. As she gets older, word retrieval is becoming less secure, and not just for names. "Mrs. . . . um, sits on the front row, grumbles. . . . Wants to borrow the church thingy. . . . Um, thing you do the grass with." This can slow things down to the point where it "more or less scuttles the whole conversation."

Thinking of what to say and how to say it with the right emphasis and timing, while listening to other people, and being ready to respond to them in time and relevantly or wittily, while looking relaxed—all this she used to do without thought or effort. Parkinson's has made her realize how precious and extraordinary chat and gossip can be. "It's shallow perhaps, but this matters to me!" It has led her to talk to herself more, and then she is unsure whether she has actually told Iain something or just thought it.

She has thought about using shorter sentences, even though she is happier in long, convoluted sentences: "That seems to be how I am in speech as in writing." When she does interject, she often finds that the moment for saying what she wants to say has passed. She wonders if others think her bored or lacking interest:

> I tried a light-hearted remark on Zoom the other day and it misfired badly. I had to email the poor woman later to apologize. I sometimes now plan in my head what I should say on Zoom, but I didn't this time, as planning it out means that I lose the right moment to speak. But I retired hurt from that experience, of unintentionally being quite rude to someone.

Conversation unfolds in real time and with elements of unpredictability as well as conventions. In contrast, her public speaking can be rehearsed and one speaker at a time is given the floor. Though those days are gone, after the worse of the pandemic she has performed readings in church with some success. During one reading she sat in the congregation with a handheld mike. She was pleased when people said afterward, "Was that really you?!"

The thing that has made her life so much richer and more enjoyable over the last three years (and it is the thirteenth since her diagnosis) has been speech therapy, "getting the proper me back." Her local group made a bid to the CEO of their local hospital to get Lee Silverman Voice Treatment (LSVT) funded, and she fronted the presentation.[7] She also wrote to ask a local MP why this was not available through the National Health Service:

Parkinson's is a variable condition, as medication kicks in and wears off, so you may not have noticed that you are losing volume (the part of the brain that listens back to one's own speech is faulty, I think) and I am articulating less well with my voice, becoming more monotonous and losing color. Couple this with a loss of facial expression, and ordinary interaction such as shopping or chatting in a meeting becomes difficult . . . and teaching becomes very tricky!

Thinking back, difficulty with the subtle combination of voice/gesture/facial expression that helps you to communicate with thirty people at once was the first sign of Parkinson's. For years I have always worked with people and used my voice; it is the hardest part since my voice and my expressive face, my communication skills, are an integral part of my personality, and I feel the real me is disappearing.[8]

Nothing came of her approach to the MP. She was fortunate, some ten years later, to find a therapist who offered her this therapy as part of a research project. LSVT therapy, like other methods, depends on repeating "aaaaaah . . ." to build up the length of time you can sustain a sound at a good volume, and deliver with pitch changes—low aaah followed by high aaah. Next was building up lists of words, phrases, sentences, and then paragraphs, all at high volume. It was very intensive: four times a week for four weeks. There was also lots of homework. Everything was recorded in numbers, including the amount of time she practiced and her voice volume. At one point Lizzie thought she gave an excellent set of "aaahs," and the therapist asked how she thought it had gone. Lizzie replied in her usual voice, "Well, I thought they were quite good," and was told off. She had to keep up the effort and volume all the time. An enthusiastic and keen student, she noticed a huge improvement in the volume and clarity of her voice, "It seemed to recalibrate the bit of my brain that notices how loudly I am speaking. But it was a conscious effort, all the time. Nothing automatic at all."

Afterward, Lizzie wrote a long note of feedback, but began it by writing,

Really, this feedback should be just a few words: "This has transformed my life." All my life I have done acting, "party pieces," learning, teaching, poetry exams, drama festivals, public speaking, things that depend on my voice and facial expression. To lose that part of me was very hard.

I was fairly loud most of my life until Parkinson's. Many people with it have difficulty assessing how loudly they are speaking, and I too was speaking far too quietly without knowing it.

I was lucky enough to get intensive speech therapy as part of a research project. Thus through the brilliant help of a speech therapist who delivered sixteen hours of therapy in her own time over four weeks, I am now much louder. This transformed my life. Without enough volume, I am sidelined in groups, not heard in shops and cafés, unreachable if there is background noise; I am the rather unwell person sitting on the edge unable to take part. But now the loud me is back and so is a most important facet of my personality. I had not realized that so much of my image of myself is bound up in my speech. Loud speech is another way of claiming territory, staking our claim to be acknowledged, establishing our position in the world. Say it loud!

One unappreciated part of the therapy followed her having to choose ten "functional phrases," said frequently. Each phrase comes from the beginning of conversations in different situations, for example, "Could you tell me which aisle the cabbage is in?" These come to mind every time she talks to a sales assistant in a shop, even if they don't sell cabbage. They remind her of the need to speak up:

> I know that I have to start as I mean to go on—if I start a conversation and the first reply is "Pardon?" my contributions crumble and it is much harder to get my volume up than if I am forceful from the start. Being heard and included are important to me,

Additional hurdles are the variability and progression of her Parkinson's. The medication becomes less effective over time, and she has more periods when "off," with her symptoms to the fore. This leads to huge variability within an hour, a day, or a year, which makes it enormously difficult for other people to help:

> Some people like to speak for me, or do my buttons up, just the same as the last time they saw me. My voice matters, to tell them succinctly what help I need. I need subtlety of voice to say what I want to say with the right words and the right emphasis.

It was easier after speech therapy when she regained some confidence in her communication skills. With COVID-19, alas, some of the complete unthinking trust in her speech that she had for most of her life, and regained after speech therapy, has dwindled:

> Before PD, I didn't have to think out beforehand what I was going to say. I do now. Poor articulation, low volume, muttering, getting muddled, are lurking, ready to pounce. I have to be on alert, always. It is no longer automatic and it is hard work.

Saying it loud then, is not without some problems. Following speech therapy "rules," she has to think about projecting her voice, about her diction, and about volume. All these take effort and contain a paradox. The aim of the speech and language therapy is to make her think about what she says. By making it more conscious, the idea is to perform it better, louder, and with more emphasis. This speech really *is* performance like she did at school all those years ago, but now is performed to deceive, to make it appear normal and automatic to others. In contrast, her Parkinsonian mumble is not effortful at all. So sometimes it's a comfortable place to retire, with just her and her voice. She makes such a good job of not allowing her illness to confine her or define her, but sometimes she just relaxes and allows it to inhabit her voice for short times, at least while talking with herself, quiet but allowed. At home she may occasionally garble her words and thoughts first thing in the morning, but that's okay. She does not attempt any complicated explanations or decision-making then.

She's learned how her voice is a dominant part of how she sees herself. With Parkinson's, how she sounds is more important to her than any of the more physical symptoms. She never dreamed that she would be combining public speaking—her new performance speech—with a more private language and speech at home. Her childhood interests in elocution and drama have not been wasted.

Get Loud

Sarah Awde, the leader of Get Loud Therapy, started work with people with Parkinson's as a course and afterward gave them exercises to do for fifteen minutes per day. But she soon realized that they would stop after a while and go backward; nothing was ever maintained. She never blamed her clients but instead the condition. Sarah told me: "Dopamine, which is deficient in Parkinson's, is involved in motivation, among many other systems. Apart from anything else, those with Parkinson's are at a disadvantage."

The condition is also progressive and degenerative, so unlike a stroke where people may improve before plateauing, those with Parkinson's do deteriorate, making it important to maintain all that they can for as long as possible. The voice becomes quieter, tails off at the end, and can become robotic and inexpressive, with little change in loudness, pitch, and inflection. As a consequence, some people with Parkinson's may be seen as both bored and boring, something not helped by coincident reduced facial expression and gesture. Gone, for instance, is the booming voice of a man seeking to dominate a room or, as we have seen, the controlled authority in a teacher's speech. Sometimes a person with Parkinson's will even presume their partner has poor hearing, so completely unaware are they that it is their own speech that has softened and become inaudible. Time and again, people with Parkinson's are simply estranged from their poverty of speech and facial expression, because either they have lost self-monitoring within the brain or the changes happened too slowly to be recognized. As a result, other people may drift away from social contact with those with Parkinson's, a source of further isolation, creating a downward drift of reduced social interaction and expression, and feeding passivity and sometimes depression.

Sarah's solution is a supercharged online speech therapy, never-ending like a Dylan tour, a community across the world backed up with a team of six made up of speech and language therapists, and song and theater coaches. She thinks everyone with the diagnosis should be referred for speech therapy from the beginning: "An ounce of prevention is worth a pound of recovery." Realistically, though, she knows her clientele will come to her when their speech becomes more of a problem. The clients have an ENT assessment first to exclude other vocal cord problems and then exercises to build up the vocal muscles and breathing. She has a forty-five-minute voice maintenance group and a twenty- to twenty-five-minute group *each day* that concentrates on voice amplitude, facial expression, and over-articulation. It is not for the faint-hearted.

Her task is to enthuse and encourage her participants. She makes them slow down for clearer, louder speech, and focuses on prosody and expression, pauses and intonation. This means their speech moves from being automatic to intentional, with them thinking about not only what is said but how it is said. She suggests that automaticity of speech involves the area of the brain affected by Parkinson's, the basal ganglia, so that making speech voluntary and intentional bypasses the area affected.

She encourages clients to express in a way that may sound too loud and feel over-expressive but really is not. She cajoles them to both act and over-act. To reassure, she uses apps to feed back their volume of sound, pitch, and variation; on Zoom she encourages them to make faces together, "lion or lemon," and to record their voices to hear them back. Role play with a theatrical actor might involve a plumber about a leak, with the mundane conversation transformed into being joyous or tragic.

After warming up, they may begin by counting or reading out loud, in particular, things that have little cognitive load to their content, allowing them to focus more on how they say it rather than what they say. Only later she encourages chatter, which requires far greater cognitive linguistic load. As we have touched on, it is unpredictable, often rapid, needs turn-taking and responses to the other, and often takes place with distractions. One has to think about one's reply before saying it, how it will be said, how the words are pronounced, at what volume, which emphasis and prosody to use, and other aspects.

Her role is energetic enough to enthuse and encourage and to help her clients overcome their Parkinsonian downregulation of expression through whichever channel. It's also remorseless. "They cannot dabble; they have to do it every time. I used to be soft tippy-toeing my way, but now I don't. It's hard, it's work; it's effort." It also demands a lifelong and continuing effort to recover and then maintain their expressivity, their self-expression—their self in expression.

The prizes can be massive. One woman told her that she's got her husband back, and another said that her husband, who would not talk to his grown-up children over the phone, had now begun to again.

She is asking for nothing less than a huge commitment to think loud, be louder, be more expressive, more varied, more everything when gripped by an illness that seduces those possessed by it to go the other way, turn in on themselves and be less in the world. Sarah refutes that Parkinson's means less control. She asks people to maximize what is left precisely in the areas most affected.

Her clients, like athletes, have to drive themselves further than they may be comfortable with. Like athletes, they train every day and, like athletes, they perform. Except that when we see an athlete we see their performance, whereas for the person with Parkinson's the aim is to not reveal the effort, but rather to appear more normal in voice and face and gesture, without advertising the work involved—a justifiable deceit.

It is through embodied expression, largely through voice, prosody, gesture, and facial expression, that we reveal ourselves and exist socially. One of the tragedies of Parkinson's, seen as a condition primarily affecting gait, tremor, and stiffness, is that many people with the condition do not realize—until it is too late—its effects on voice and on communication.

There is one downside. Over months and years, many decline as Parkinson's grip tightens. Some clients have died, which is hard for those who have worked with them. But the clients are well aware of their prognosis, and the group has a strong mutual bond and a shared implicit optimism to go on while they can. They know there will be a dying of the light, but rather than going gently or meekly into that good night, they are doing their best to keep loud and to remain expressive.

9 In My Head I Talk Slowly: Cerebral Palsy

Cerebral palsy (CP) is caused by abnormal development of the brain or injury to it, often during or soon after birth, and affects movement, balance, and/or posture. Some people need full, lifelong care, while others have only slight problems with gait or balance. It may be divided into several types. The commonest leads to stiffness in the limbs and trunk, making movement difficult, caused by damage in the descending control by the brain. It can affect the legs only (diplegia, paralysis; or diparesis, weakness), or the arm or leg on one side (hemiplegia or hemiparesis). This form of cerebral palsy affects the majority of people with the condition, 80 percent or so. In dyskinetic CP, the difficulty is uncontrolled rather than reduced movement: the affected body part writhes slowly or makes unbidden rapid jerking movements, which may affect walking or sitting. It can affect the arms and/or legs, or head and neck control, so-called oro-facial-pharyngeal CP, affecting speech and swallowing. Ataxic CP, in contrast, affects primarily balance and posture. These latter two types each affect less than 3 percent of those with CP, with the remaining number comprising those with a mixed pattern. It has been estimated that CP affects 0.3 percent of live births, meaning there are around 750,000 people with the condition in the United States.[1]

This chapter and chapters 10 and 11 consider the experience of adults living with cerebral palsy, first within a community in Denmark with various degrees of severity but who all have problems with speech, and then that of a Japanese philosopher whose insights and eloquence have been refined by her chosen profession.

Jacob and Martin are both associated with Kristian Martiny's Enact:Lab in Copenhagen,[2] which exists to encourage social change in relation to impairment. Kristian undertook a doctorate studying those with cerebral palsy but is also active in the media. He founded Enact:Lab to use whichever tools are available to make society more tolerant and accepting of diversity. Jacob and Martin, who both live with cerebral palsy and speech difficulties, collaborate with Enact:Lab, while Jacob also runs his own company (Nossell and Co.), a consultancy to build bridges between society and disability through different projects and communication. Martin also works for it. Together we proposed having a day where people with cerebral palsy might come together to discuss their experiences and problems with speech, and agreed it should be run by Jacob and Martin.[3]

They began by defining the term "voice" etymologically and then explored how our "voice" also holds a psychological meaning:

> Our voice expresses our mood and mindset. Our words and the tone of our voice express our personality. We raise the voice when we're happy and lower it when we're sad. The pronunciation of our words, and the way we relax and tighten the muscles in the diaphragm and the vocal cords, is essential in producing the tones and overtones of our voice. We also read and we think with an *inner voice*. As such speech is both social and communicative but deeply cognitive and might be said to be the carrier of linguistic consciousness.

Jacob and Martin's concerns are with these biological and psychological aspects, to try to understand the social aspects of the voice. Living with a speech impairment themselves, they also realize that having a voice means nothing without a listener. As Jacob said, "We can scream as much as we want, but our voices and our personalities have no agency until received and understood and through interaction with others; conversation exists not in speech but between people mutually engaged in dialogue."

They wanted to explore how social interaction and dialogue develop in those with problems with communication. Jacob and Martin were also interested in social inclusion and how voice affects others' perception of a person in relation to competence and intelligence, so relevant to those with CP. These aims were quite ambitious for an interview. I was reminded of an astronaut friend who is often asked what it is like in space. She replies by describing how she sleeps, eats, moves, cleans her teeth, and—the one we are most intrigued by—how she goes to the loo. The big questions are often best crept up on by exploring the minutiae from the bottom up. Fortunately, despite their elevated aims, Jacob and Martin's set of prepared questions were more granular and lower-level.

They arranged to meet a group of people with speech problems associated with cerebral palsy. Participants were recruited through the Danish Enact:Lab's network (courtesy of Kristian Martiny) and through relevant disability organizations. The interviews were held at the house of Disabled People's Organisations Denmark (DPOD) in Høje Taastrup, a suburb of Copenhagen. The aim was to have as interactive a conversation as possible, though it was anticipated that with the four interviewees and two moderators speech might flow slowly.

Neither Jacob nor Martin had tried this before. Their own speech impairments would certainly alter the dynamic. It would be slow, though we agreed their presence should allow more equality and enable a feeling of community and safety.[4] To avoid disrupting the conversation, I sat in another room watching on a laptop screen with Kristian, who kindly translated.[5]

In our letter of invitation we had introduced ourselves and our overlapping interests, but we began with me repeating my purpose and with Martin and Jacob reiterating theirs. They had prepared a series of questions and planned to progress through them.

> Are there situations when your speech impairment is more problematic?
> How does that affect you?
> Are there situations when you cannot make yourself heard clearly?
> Is fatigue a factor in your speech?
> Are there situations where you have felt different or excluded?
> Can you compare conversations with people with and without a speech impairment?
> How does your speech impairment influence others' attitudes toward you?
> Have you had trouble showing others what sort of person you are?

Jacob, who has considerable media experience on TV and radio in Denmark, has a mild hemiparesis and speech impairment. Martin can walk but often uses a wheelchair; he has involuntary movements and spasms, which make speech difficult. His body writhes and jerks when he tries to vocalize: intentional movements frequently trigger unintended ones, as is often the case in CP.

Four interviewees attended. Mikkel, in a wheelchair, is a singer and actor: "I have a natural interest in how voice works and how to express myself in speech." Vibeke also lives from a wheelchair and was also very engaged: "It is so important to focus on communication like this. I wonder why no one has done it before. It's fucking good." Emilie also lives from a chair and like the others arrived with her personal assistant (PA). Susie walks in with a spastic gait and her PA. She was in her fifties. She explained,

> I always have a speech impediment, but in the last ten years it has got worse. I am very interested in this project. Over the last few years my stutter has worsened. My air, lungs, tongue, everything is a bit worse. I always had a speech problem but now it is speech, swallow, and stutter.

As we have suggested, when one thinks of a person with CP, it is often a child, but children grow into adults and then those adults age and may then run into more problems. Those with CP and other impairments have the usual diseases of old age—arthritis, for example—but can also become progressively weaker. They can be helped by training weakened muscles as best as possible, though this itself can be an effort.

Jacob introduced the central topic: how voice affects social interaction in CP. Martin then talked of his inner voice, and what he hears when he talks:

> I don't hear with a disability at all. What is in my mind is what I hear. Then, when I hear myself on tape: "Holy shit." What a difference between what I hear, what I think I am sounding like, and what comes out for everyone else. My PA knows when I am emotional, angry, or whatever from my voice, but this may not be so apparent to others. When stressed, my speech deteriorates markedly; when pressured, it becomes difficult.

Susie's concern was not with her speech impairment due to CP, but with an increasing stutter. "As a result, I am more annoyed than I was in my twenties and thirties." Taking up Martin's point, Emilie mentioned that she too has a fluent inner voice: "I did not know I spoke differently till I was aged six. It's a problem now when I am angry or drunk, though sometimes I think I am more fluent then." Both Mikkel and Vibeke agreed that their inner voices were fine and that how they were able to talk depended so much on how relaxed they felt in the company. Vibeke mentioned how her "PA says how different I sound with my parents or with others."

We all speak differently according to the audience and how relaxed we feel, but it seems evident that for those with CP this is a huge concern. Emilie once more:

> The intensity of the situation and the feeling and the way everything is affected seems greater with CP. My voice, my hands—it's everything. I feel my voice tense, utterance, stutterance—it's all worse with spasticity in my hands. I feel it now more than when younger.

Jacob asked whether tenseness was a problem. They agreed that it was worse when they were out with unfamiliar people and with larger groups, and, unsurprisingly, when speaking a foreign language. Being in Denmark, the latter is less uncommon than, say, in the United States or United Kingdom. They were always aware too that their speech was discriminated against. Mikkel: "I once hit another car, phoned the police who assumed I'd been drinking. It is much more difficult to be understood on the phone."

Martin commented on not being understood: "In company I prefer to be asked to repeat than for them to just say 'Yes' and move on. Anyway, explaining breaks the ice a bit. I will say the same thing using different words till they understand, again and again." Emilie said that sometimes she is not understood despite repeating something ten times. The uncertainty leads to insecurity: "I understand people don't want to say, 'What did you say?' But just to say, 'Yes,' with no understanding at all is just so disrespectful. Ask again and again, please. It must be difficult for people; them being uncomfortable makes us uncomfortable even more so." Vibeke mentioned that as soon as some people hear her voice she feels prejudice and bias. Some people just look away and shut up.[6]

Jacob asked about use of vocabulary: "Everyone knows simple words, like 'Shut up,' but longer ones are more difficult to use. Do you think about which words you use?" All agreed that they did, according to "whom I am speaking to and how, and the situation." Words and sentences required thought before speech.

Vikebe had considered this a lot: "Not so much about my speech problem, but about the audience and its likely reception." Susie also had her stutter to negotiate with: "One word can be blocked, so I have to find another. Maybe not the one I want but one my spasm and stutter will allow." Even without a stutter, it happened to others too. It was also a problem with new PAs, who need to be taught to tune in: "Initially, they can be uncomfortable and I have to make them calm and move them from notions and words they know to new ones."

Martin asked how they projected themselves and avoided being unseen and ignored. Most had had the experience of a shopkeeper talking to the PA rather than to them. Vikebe mentioned that in her local bakery the counter was so high that she could not be seen. Emilie found it so much harder in larger groups: "Our voices are difficult and we cannot increase the volume." Susie mentioned that sometimes across a table it's difficult to make herself heard. "If the other cannot hear me and there is no communication, then I just cut it. Sometimes the other is not honest enough to say they cannot understand. My own child has been asked if she understands me—of course she does. It depends how you listen; if you are used to it, then it's not a problem."

Mikkel was more assertive: "In our culture when you have something to say, you just say it. You don't get a number for your turn or pass a conch shell. So I will talk over another. I say, 'Just breathe, and I will speak.' Naughty I know, but why shouldn't we be?" Indeed, why should people with impairments have to behave the whole time? For Vibeke, noise and the presence of lots of people mean she has to pull herself together or she may panic before saying anything. Emilie mentioned how normal her life is with parents and sister compared with other groups. Her speech and swallowing have worsened later in life, with more choking on spit: "My normal is changing."

Jacob asked them about people taking over and the role of power in conversation. Susie focused more on her stutter: "A huge additional difference. People guess what I want to say; it annoys the shit out of me." Vikebe agreed: "It depends on my mood. On a bad day I get fucking angry. You get some people who think they understand and take over. So annoying." Susie said, "Many times I have not been taken seriously; [with the assumption] that nothing in my head or brain works. If given space and time people are surprised [at how I can express myself]. Many assume we have no knowledge of anything." They moved on to discuss whether it was better now or when they were young. Some thought there was less time now and that people were more aware of the smallest imperfections and so less forgiving. However, in his empirical work Kristian has found the opposite, with more openness toward those with impairment. The worry was that older people with CP may become increasingly marginalized.

The conversation flowed, with everyone contributing as they wanted, and though some of Jacob and Martin's questions went unanswered, they had explored much over a two-hour period. Observing from another room, their various impairments of speech were irrelevant once a rhythm and pace to the discussion had emerged. To see someone with CP speak is extraordinary. What for most of us tumbles out with little or no thought, for them appears to be a continual duel between their intention and their spasms. We can see the neck and head moving, and with these their bodies and arms writhing continually, apparently without reason, but we cannot see the vocal muscles, diaphragm, oropharynx, and mouth and nasal tract control that underpins speech. To make sounds and words, let alone intonation and expression, while battling such spasms and incoordination and also being sensitive to others' expectations, must be exhausting.

Over lunch I tuned more into Jacob's speech, with its own cadence and rhythm.[7] He has to concentrate to pronounce and say the words correctly—and with the right rhythm and tone. "There is also the question of fatigue when I talk. I need small pauses. New words are difficult too. Every new word must be hard coded in some way. I also often 'forget' words when I write because I think faster than I write."

I asked about the mental effort of turning thoughts into pronounceable words, and the effort of concentrating on making the words and sentences and listening to others and formulating a reply, even without fatigue in the vocal cords and throat. "The effort is a combination of all. It's hard to explain. It is like my sentences have to be transcoded twice, one time with what I want to say, and a second time with what is possible to say." He is also aware of thoughts he cannot express:

All the time. In my head I speak freely. Thoughts to words are a constant "transcoding" process with different layers of transcoding to do before there is a proper output. That is why I rely on second-guessing others to reduce the effort. I can be a pain to talk with—because I second-guess a lot to get a head start. I am good at it, but sometimes I get it wrong. It is a question of complexity.

I use all my energy when I am out or at work—and I used to collapse when I got home or with good friends. I have a problem now—now that I am a dad—I realize that I cannot be silent anymore at home. I have to speak with my son, even though I am very tired sometimes.

After a morning spent with half a dozen people with CP exploring their voices, it was good to hear of Jacob's home life. Welcome to parenthood, I thought. After we all trooped off to have lunch, Martin and Jacob and I reassembled to meet a person with more severe CP, but before that I had the chance to talk with Martin about his experiences.

After lunch I sat with Martin. As a teenager he remembers watching a video from when his little sister was born and he was a young child:

> "What am I saying, mom?" I asked. We were together in the living-room watching the video. My mom understood what I said, but I couldn't understand my own speech as a boy. It was weird! It sounded so differently in my head compared to my voice on the video.[1]

Now in his thirties, his cerebral palsy involves his head, neck, and body, with spasms, reduced coordination, and tremor when he tries to move. He can walk but finds a wheelchair easier. He writes using a large type pad with some adaptations. Writing is easier than speech, which is usually in single-word expulsions or short phrases. While he may be initially difficult to understand, one soon tunes into this, though usually he has a PA who can help. He first became aware of his voice being a problem as a teenager when his speech became less understandable:[2]

> It was embarrassing. My voice became associated with shame. My mother always got frustrated when I didn't concentrate enough on speaking clearly. The speech therapist said that my adolescent speech was normal because of physical changes to the speech muscles. I remember having to practice to avoid drooling as a child.

His school was unique in having a separate section for children with disabilities (mostly with forms of cerebral palsy). He started preschool and through all his nine years there, though he had some activities with others with CP (physiotherapy, speech therapy, and occupational therapy), he also attended an ordinary class as the only child with disability, with a supporting teacher during some lessons. As he progressed from the lower second grades and up (years seven to nine), he needed less support and was mostly on his own. His parents still had to argue with the schoolmaster and local officials every year whether or not he was "good enough" to continue. Some parents wanted him out, as did an old conservative master. But Martin was not without support. At times his teacher put her job at stake to keep him in her class. He was not aware of this at the time, though he was aware of the need to speak well enough to stay "single-integrated" in a normal school class.

His memories of school are not all about problems. During primary school each year all disabled pupils went to Norway for a week of winter camp. It was a huge effort for the teachers and assistants, and indeed to raise the funds. It started with ten people but soon swelled to thirty or forty. One time Martin was sitting in the restaurant on the ship from Copenhagen to Oslo. There was lots of excited talk and music around, when suddenly from the kitchen there was the sound of lots of plates falling and smashing. Martin turned and spoke to his special class teacher, but the background noise was too loud for him to understand. Martin tried repeating, but it remained unintelligible. "Can someone please help me understand what Martin says?" No luck. "Martin—let's walk out of the restaurant and find a quiet spot." He pushed Martin through the long corridors on the ship until finally they found somewhere quiet. The teacher squatted down in front of Martin. "Okay Martin, let's hear what you were saying—I really want to understand." Martin repeated what he said once more: "Oops!" The teacher asked if that was all. "Did we really go all this way only for this?" "Yes."

What is visible when Martin talks is effort, with dystonic movements of the head and neck interfering as best they can. Seeing his neck and body misbehaving, one can only imagine the spasm and incoordination within the larynx. It turns out the problem can begin even earlier in the vocal apparatus:

> I've learned that anxiety is frequent in people with dystonia (and maybe also CP in general), which makes perfect sense when your body doesn't react to your intentions. I also think that the dystonic/hypertonic movements in my diaphragm made me extra anxious, especially in social settings. This tremor made me feel that I couldn't even control my own breathing. It affected my overall well-being and self-esteem of course—and my courage to talk.

Not just his head and neck or even the vocal muscles; he cannot even rely on expelling air as he would like via the diaphragm. He has to negotiate with the whole of his vocal tract to speak, word by word, effort by effort. His mum used to be upset if he didn't concentrate enough when speaking.

"Focus on one word at the time." The problem is that during everyday talk, he cannot think of what to say and at the same time concentrate on trying to overcome the spasticity:

> The uncontrollable movements always come when I'm aware of my body, or if I'm not in "balance" with myself and the world around me mentally. For many years I had this shameful feeling of not being able to speak as well as I was supposed to. I would be left out from group talk, because my spasticity increased when I wanted to butt in, making it impossible. Sometimes I would think that maybe what I wanted to say wasn't interesting, or they probably wouldn't understand me anyway.

He was also aware that his dystonia may have been misperceived by others as a mental disability. He felt "wrong" or "out of place," a feeling increased by his lack of fluency in speech:

> There was also a vaguer but more pervasive sense of being "out of the world that others lived in"—which also had to do with my cognition. During my master's thesis, I interviewed a neuropsychologist who told me that people with problems communicating (mostly due to a trauma later in life) can feel detached from life, or "three steps away" from others.
>
> It took me more than thirty years to realize that not only am I speaking slowly, but there's also a thing about energy consumption. It is simply physically exhausting to speak. I guess my body has always known that this was the case but my mind wouldn't accept it. And logically of course, when it's exhausting to speak and to small talk, you automatically just shut up.

Most of us would find it difficult to imagine talking being exhausting; it just happens. Martin has to be aware of the capricious, almost conspiratorial quality of his body's movements. They can be relatively quiet and disinterested until he tries to express a word or sentence, at which point his neck and vocal tract awake to frustrate him with bucking, seemingly random, and at times malevolent co-contractions:

> I've been training and training and training to be aware of this when speaking, but it doesn't come naturally. My intentions are mixed with spasms unpredictably the whole time. Spasticity—and dystonia especially—affect one's mental state of mind, and vice versa. I can have lots of good intentions (of being relaxed) but my body does the opposite.

At least it is easier now than when he was young. He understands his difficulties and crucially he knows his own worth and feels more self-confident. "No one tells me that I'm not good enough anymore." An academic education and recognition of his intellectual abilities has given him not only knowledge and insight, but the confidence so lacking in child-hood and the ability to be more "in sync" with how others see him. He still meets situations where he is presumed to have lower mental capacity but now has a group of people who see through his disability and ignore those who stare.

Though finding the words during speech comes more naturally now than before university, deep conversations have been very rare in his life. "This has been a huge deprivation for many years. I felt I couldn't get close to people and I missed having close friends." He was fortunate to go away to a boarding school for a year when he was fifteen, which focused on improving those with disabilities. He made some close friends there, boosting his self-esteem. That made returning to normal high school after the year away a huge anti-climax. Despite this, he went to university and gained his degree:

> The biggest challenge for me still is getting to know new people because of the presumptions and stigma associated with my speech. Like everyone but far more I still feel anxious in a new place with new people. Informal interaction can be really difficult. I sometimes feel totally alienated because my situation and my body are so different. I never know how to begin a conversation.

He often has thoughts he cannot express:

> Loneliness is a premise. I often feel different or wrong or out of place in social settings. Up until my thirties, I couldn't figure out why there was this mismatch between my sense of self and other people's perception of me. I knew that I was smart and reflective—they knew it too—but still there was this wall between me and the world around me. I guess it has a lot to do with sexuality and not being recognized as a sexual being.
>
> I have a few close friends with whom I can express lots of things. I think what is most difficult for others to understand is the experience of being so dependent on others, my personal assistants. Even some people with a milder CP are not always aware of this.

In his thirties he saw a psychiatrist, with depressive thoughts as to what his life was about (in his words: "what the fuck it was about"). He still sees them now and then—and they have helped him a lot. He realizes that he was not the root of his problems. Being recognized—seen, validated—by others, including as a sexual, attractive individual—as he sees himself—is crucial. He still battles with how sex and disability are taboo to talk about. "This made it very difficult to find a partner and have experiences. Like a bad circle."

Education has had a huge impact on enlarging his vocabulary. "I think I was slow in getting a suitable vocabulary because I was not used to small talk. Maybe until my education my world was not so big and my vocabulary limited. Writing always gave me better opportunities to express myself, not least because it proceeds at a rate that suits my thoughts better than using my voice." Martin writes using an oversized adapted keyboard. Despite his arm and hand movements being large and not always under complete control, he has a knack of being able to land a hand or finger on the correct key as his hand flies past. This is easier than controlling the vocal muscles during speech, though neither is without uncertainty or risk. At least writing can be seen and edited, giving him more control than speech, which is one-off, performative, and not correctable. That may reduce his anxiety and spasm during writing.

Timing is always problematic because of the spasm and tremor. Any movement is imprecise and shaky; intention is translated into action with varying degrees of imprecision. Fatigue is something Martin has become more aware of, whether speaking in his native Danish or the more difficult English, where he also has to concentrate on the word and holding it in his mind as mind negotiates its pronunciation with his body:

> I only became aware of the importance of fatigue recently. My mother knew about me having to use more energy than others especially when speaking. "I can hear from your voice that you're tired," she might say. But fatigue is a new concept for me. My understanding is that it is both physical and psychological. I've become more aware of its symptoms. It affects my short-term memory (which is not the best after all) and my desire to be social.

As overcoming or at least negotiating with the co-contractions, spasm, and tremor in order to speak is physically tiring, so is it a continual mental effort, and sometimes this is too much. The balance between reward and frustration is also hugely dependent on those with whom he is engaging. This can be the case for all of us, but for those for whom the effort to express something is so complex, and for whom their difference is so visible, this becomes crucial.

Turn-taking remains difficult. It depends on the situation and how much is at stake. Small talk can be okay in smaller groups if he can be understood properly. But whenever he feels judged, say, in social groups with an elitist vibe, he feels excluded and insecure and then becomes introspective, fearful of being unable to live up to expectations:

I tend to withdraw and resign myself to silence when I feel too much out of place, or when I feel that people are too insecure to get in touch with me, or whenever the situation doesn't allow me to break through prejudices. I've learned to not take it too personally but it will probably always be annoying. Sometimes I spot a person with the right amount of empathy and patience, and it makes my whole world better.

After talking with Martin, another participant arrived, driving his wheelchair into the room. He was as gaunt as a marathon runner, though he cannot run or walk or stand. It was as much as Jeppe could do to stay in the chair given his spasm and involuntary movements. His legs were out in front of him, resting (or tethered) on foot plates. He sat behind a Perspex table with a red knob that allowed him to steer, and handholds that he gripped, seemingly to prevent his arms flying off unbidden. One eye turned inward, and while the other one was useful, this was only as long as he compensated for the unpredictable, writhing movements of his neck and head. He was gaunt because eating was a problem, risked choking, and because his movements and spasms impose a huge metabolic load. He was wearing a singlet, reinforcing the athlete analogy, despite it being winter in Denmark and freezing outside. We were warmed by our sweaters; his spasms and movements kept him warm, like a boxer. Behind his head was a support to prevent his neck from over-extension backward. It worked but I saw that he had one cauliflower ear. In the United Kingdom these were usually seen in rugby players who had buried themselves in a scrum and suffered repeated bruising and bleeding in the external ear, or pinna, leading to permanent swelling. I presumed Jeppe's spasms have, over the years, bashed his ears into the headboard.

His PA followed and explained that Jeppe used three ways to communicate. Usually his PA interprets but he also has an eye tracker, which allows him to direct his gaze to an alphabet and then accept a letter and build these into words and sentences. More recently, he has had some singing lessons, and these allowed him some vocalizations, more than he imagined though still too difficult to use with us.

As in the morning, I watched Jacob and Martin interview Jeppe from another room, with Kristian and his research fellow Emilie translating. Before that I explained my background in neurology and how I became involved with the voice when collaborating with people with vocal cord paralysis. I mentioned that they had made me understand the importance of the voice and that I had been exploring this by meeting those for whom the voice was a problem. I explained how I knew Kristian and how grateful I was to him and his team, and especially to Jacob and Martin, for their interest. As I spoke, Jeppe and his PA sat next to each other. I spoke clearly and looked for the most part at Jeppe, knowing how important it was to address him. It was not straightforward to do this, though. His PA looked intently at me and responded to what I said, whereas Jeppe's gaze was difficult to maintain. His presence was undeniable, but focusing on his body was not easy.

Jeppe introduced himself to us. He turned his head to the left and his PA began a recitation that soon became familiar, "1 2 GHI, 1 2 GHIJKL, 1 2 GHI, 1 2 3 4 . . ."

Their system of communication became clear as we all looked on. His PA recited a number from 1 to 4, which referred to quarters of the alphabet. At the right number, he then moved to the alphabet and began saying letters, quickly but distinctly, to give Jeppe a chance to interrupt at the right one. How Jeppe showed this we could not spot initially. During a break, his PA says the "tell" can be a small drop of the head or, if Jeppe is tired, his eyebrows rose. We found it difficult see these, since all were superimposed on and interfered with by the spasms.

Alphabet

1 A

1 A B

1 A B C

1 A B C D

1 A B C D E

1 A B C D E F

2 G

2 G H

2 G H I

2 G H I J

2 G H I J K

2 G H I J K L

3 M

3 M N

3 M N O

3 M N O P

3 M N O P R

3 M N O P R S

4 T

4 T U

4 T U V

4 T U V Y

4 T U V Y Æ

4 T U V Y Æ Ø

4 T U V Y Æ Ø Å[3]

> 1 2 GHI 1 2 GHIJKL 1 2 GHI 1 2 3 4 TUV 1 ABCDE 1 2 GHI 1 2 3 MN 1 2 3 4 T
> 1 2 GH 1 ABCDE 1 2 3 MNOPRS 1 2 3 4 TU 1 AB 1 2 3 4 TU 1 2 3 MNOPR
> 1 AB 1 2 3 MNOPRS
> "I live in the suburbs."

> 1 2 GHI 1 A 1 2 3 M 1 2 3 4 T 1 2 GH 1 2 GHI 1 2 3 MNOPR 1 2 3 4 T 1 2 3 4
> TUVY 1 2 3 4 T 1 2 3 4 TUV[W] 1 2 3 MNO 1 2 3 4 TUVY 1 ABCDE 1 A 1 2
> 3 MNOPR, 1 2 3 MNOPRS 1 2 3 MNO 1 2 GHIJKL 1 ABCD
> "I am thirty-two years old."

This was Jeppe's introduction for me and the team. Being with Jacob and Martin in one room with their PAs, and with Kristian, Emilie, and myself in another, was quite an anxiety-inducing environment; however, Jeppe seemed relatively relaxed.

I was spellbound as his PA ran through the litany of letters and numbers, searching out Jeppe's words and sentences. Jeppe had to formulate a sentence in his mind, then remember it as he reduced each word to letters. His PA had to pick up these letters from Jeppe's tells and then remember the word and sentence as it built up and then relay it to us. Only then did Kristian translate these for me. One of us measured an answer as taking eight minutes, but time was relative. We were inside the action unfolding between Jeppe and his PA. In another room in another language, I just looked at the screen as Jeppe's speech unfolded, letter by letter, into sentences and meaning.

Jacob asked a question. "Is some communication more difficult than others?"

1 2 GHI 1 2 3 4 T 1 ABCD 1 ABCDE 1 2 3 MNOP 1 ABCDE 1 2 3 MN 1 ABCD,
1 2 3 MNOPRS
"It depends . . ."
1 2 3 4 TUV[W] 1 2 GH 1 2 3 MNO 2 GHI 1 A 1 2 3 M

"who I am with," his PA said, predicting the last word. With friends Jeppe has to think less. Social relations and conversations with family and friends were longer and deeper. With friends and family, he can also express himself in longer, deeper, more personal ways, but outside he keeps it simple and short.

From thought into sentences into words and then letters, relayed via the laborious alphabet system to be built back up into words and sentences by his PA might sound slow and boring, but we watched enthralled. Jeppe had his head turned to the left to face his PA while his left hand gripped the handlebars to stop it upsetting things, and his right hand was permanently, it seemed, held with bent elbow up to his face, fingers extended as he looked at his PA. I tried to look for the tell when the right letter had been reached. It was not easy for someone unfamiliar with the process. His PA admitted that he sometimes forgets what the words were; after all, each sentence takes minutes to emerge. When Jeppe talks more philosophically, his PA may even write it all down as he goes.

Vygotsky wrote that consciousness needs language and shared expression. Jeppe's communication, our perception of his consciousness, his self, is through his PA and through the laborious, repeated alphabet. He has his alphabet tattooed onto his left forearm, a tattoo of his way into the world through language.

Jeppe said that when he says something it sounded normal in his head, but then as he expressed it, he knew this was not how others heard him. To translate thoughts into expression was always slow and required him to slow his thoughts, holding a thought—even a simple one—for as long as it took for his PA to reconstruct it. At home he tells jokes and his family laughs, even though it may take ten minutes for the punch line.

He has found some PAs are better than others; one who has known him for years can almost pick up his feelings from his eye. Others try to predict what he wanted to say, but he always insists they use the right word each time.

His PA expresses Jeppe's words and as he does so he gestures too. Whose gestures are they? I thought of a disabled athletic club in Copenhagen,

Figure 10.1
Photo with permission of Jeppe Forchhammer (who claims copyright).

some of whose members with cerebral palsy entered marathons. Pushed along in their chair, they still feel part of it and still feel some sense of agency toward it. There is a film of a Japanese farmer with severe motor neuron disease. Though immobile and on a breathing machine, he was lifted into the driver's seat of a large combine harvester, which then moved up and down a field guided by GPS. He felt he was still farming and did so with a smile on his face.

Jacob asked, "Are there some situations which are more difficult than others?"

> 1 2 3 MNOP 1 2 3 4 TU 1 AB 1 2 GHIJKL 1 2 GHI 1 ABC 1 2 3 MNOPRS 1 2 3
> MNOP 1 A 1 ABC . . .
> "Public spaces."

Jeppe's PA is an expert predictor and sometimes does short cuts. If wrong, Jeppe brings him back to say exactly what he wants to say.

> 1 AB 1 ABCDE 1 ABC 1 A 1 2 3 4 TU 1 2 3 MNOPRS 1 ABCDE 1 2 3 MNOP 1
> ABCDE 1 2 3 MNO 1 2 3 MNOP 1 2 GHIJKL 1 ABCDE 1 ABCD 1 2 3 MNO

1 2 3 MN 1 2 3 4 T 1 2 3 4 TU 1 2 3 MN 1 ABCD 1 ABCDE 1 2 3 MNOPR 1
2 3 MNOPRS 1 2 3 4 T 1 A 1 2 3 MN 1 ABCD 1 2 3 4 TUV[W] 1 2 GH 1 2 3
MNO 1 2 GHI 1 A 1 2 3 M 1 A 1 2 3 MN 1 ABCD 1 2 GH 1 A 1 2 3 4 TUV 1
ABCDE 1 2 3 4 T 1 2 GH 1 ABCDE 1 2 GHI 1 2 3 MNOPR 1 2 3 MNO 1 2 3 4
TUV[W] 1 2 3 MN 1 ABCDE 1 2 3 4 TUV[X] 1 2 3 MNOP 1 ABCDE 1 ABC 1
2 3 4 T 1 A 1 2 3 4 T 1 2 GHI 1 2 3 MNO 1 2 3 MN 1 2 3 MNOPRS
"Because people don't understand who I am and have their own expectations."

His PA explained that often people see the chair and a disability rather than
a person and presume a cognitive disability as well.

Jacob continued. "What are the major challenges?"

1 2 3 MNO 1 2 3 MNOPR 1 ABCD 1 ABCDE 1 2 3 MNOPR 1 2 GHI 1 2 3 MN 1
2 G 1 ABCDEF 1 2 3 MNO 1 2 3 MNO 1 ABCD
"Ordering food."

No high-level philosophical stuff here. I was picking up on Jeppe's sense of
humor.

1 ABC 1 2 3 MNO 1 2 3 MN 1 ABC 1 ABCDE 1 2 3 MNOPR 1 2 3 4 T 1 2 3
MNOPRS 1 2 3 MNOPRS 1 2 GH 1 2 3 MNO 1 2 3 4 TUV[W] 1 2 3 MNOPRS
1 ABCDEF 1 ABCDE 1 2 3 MNOPRS 1 2 3 4 T 1 2 GHI 1 2 3 4 TUV 1 A 1 2
GHIJKL 1 2 3 MNOPRS
"Concerts, shows, festivals."

He was due to open a congress for the International Society for Augmenta-
tive and Alternative Communication (ISAAC). He plans to talk about sex.

Jacob asked, "How do you communicate every day?"

1 2 GHI 1 ABCD 1 2 3 MNO 1 2 3 MN 1 2 3 4 T 1 2 3 MNOPR 1 ABCDE 1 A 1 2
GHIJKL 1 2 GHIJKL 4 TUVY 4 TU 1 2 3 MN 1 ABCD 1 ABCDE 1 2 3 MNOPR
1 2 3 MNOPRS 1 2 3 4 T 1 A 1 2 3 MN 1 ABCD 1 2 3 4 T 1 2 GH 1 ABCDE
1 2 3 MNOP[Q] 1 2 3 4 TU 1 ABCDE 1 2 3 MNOPRS 1 2 3 4 T 1 2 GHI 1 2
3 MNO 1 2 3 MN
"I don't really understand the question."

(I was unclear whether this meant he did not understand Jacob, or get the
meaning of the question.) Jacob repeats it slightly differently: "How much
do you communicate normally?"

1 2 ABCD 1 2 GHI 1 ABCDEF 1 ABCDEF 1 ABCDE 1 2 3 MNOPR 1 ABCDE 1 2 3
MN 1 2 3 4 T 1 2 3 MNOP 1 2 GHIJKL 1 A 1 2 3 4 T 1 ABCDEF 1 2 3 MNO 1
2 3 MNOPR 1 2 3 M 1 2 3 MNOPRS 1 2 ABCD 1 2 GHI 1 ABCDEF 1 ABCDEF
1 ABCDE 1 2 3 MNOPR 1 ABCDE 1 2 3 MN 1 2 3 4 T 1 2 GHIJKL 1 ABCDE
1 2 3 4 TUV 1 ABCDE 1 2 GHIJKL 1 2 3 MNOPRS
"different platforms, different levels."

His PA mentioned that through his computer he used Facebook. Jacob asked which was easier, through Facebook or his PA. Jeppe answered and again I wondered about his sense of humor.

> 1 2 GHI 1 2 3 4 T 1 ABCD 1 ABCDE 1 2 3 MNOP 1 ABCDE 1 2 3 MN 1 ABCD 1 2 3 MNOPRS 1 2 3 MNO 1 2 3 MN 1 2 3 4 T 1 2 GH 1 ABCDE 1 2 3 MNOPRS 1 2 GHI 1 2 3 4 T 1 2 3 4 TU 1 A 1 2 3 4 T 1 2 GHI 1 2 3 MNO 1 2 3 MN 1 2 GHI 1 ABCDEF 1 2 3 MN 1 2 3 MNO 1 2 3 4 T 1 ABC 1 2 3 MNO 1 2 3 M 1 2 3 MNOP 1 2 GHIJKL 1 ABCDE 1 2 3 4 TUV[X] 1 2 GHI 1 2 3 4 TU 1 2 3 MNOPRS 1 ABCDE 1 2 3 MNOP 1 A 1 2 GHI 1 ABCDEF 1 ABC 1 2 3 MNO 1 2 3 M 1 2 3 MNOP 1 2 GHIJKL 1 ABCDE 1 2 3 4 TUVW[X] 1 2 3 4 T 1 2 GH 1 ABCDE 1 2 3 MN 1 2 GHI 1 2 3 4 TU 1 2 3 MNOPRS 1 ABCDE 1 2 3 4 T 1 2 GH 1 ABCDE 1 ABC 1 2 3 MNO 1 2 3 M 1 2 3 MNOP 1 2 3 4 TU 1 2 3 4 T 1 ABCDE 1 2 3 MNOPR

"It depends on the situation. If not complex I use PA, if complex then I use the computer."

Jacob continued with questions, "What is important for an enjoyable conversation?" Seeing the effort Jeppe put into all language, the idea that it might be pleasurable was incongruous, though I knew better than to see from my own perspective.

> 1 2 3 M 1 2 3 4 TUVY 1 2 3 MNOP 1 A 1 2 3 MNOPR 1 2 3 4 T 1 2 3 MN 1 ABCDE 1 2 3 MNOPR 1 2 3 MN 1 ABCDE 1 ABCDE 1 ABCD 1 2 3 MNOPRS 1 2 3 4 T 1 2 3 MNO 1 AB 1 ABCDE 1 2 G 1 2 3 MNO 1 2 3 MNO 1 ABCD 1 A 1 2 3 4 T 1 2 3 MNOPR 1 ABCDE 1 A 1 ABCD 1 2 GHI 1 2 3 MN 1 2 G 1 AB 1 2 3 MNO 1 ABCD 1 2 3 4 TUVY 1 2 GHIJKL 1 A 1 2 3 MN 1 2 G 1 2 3 4 TU 1 A 1 2 G 1 ABCDE.

"My partner needs to be good at reading body language."

To my untrained eye much of this was imposed by his spasm and the CP, but obviously I needed to tune in better.

> 1 2 3 GHI 1 2 3 4 T 1 2 GHI 1 2 3 MNOPRS 1 2 GHI 1 2 3 M 1 2 3 MNOP 1 2 3 MNO 1 2 3 MNOPR 1 2 3 4 T 1 A 1 2 3 MN 1 2 3 4 T 1 2 3 4 T 1 2 3 MNO 1 2 3 GHIJK 1 ABCDE 1 ABCDE 1 2 3 MNOP 1 ABCDE 1 2 3 4 TUVY 1 ABCDE 1 ABC 1 2 3 MNO 1 2 3 MN 1 2 3 4 T 1 A 1 ABC 1 2 3 4 T 1 2 3 4 TUV[W] 1 2 GHI 1 2 3 4 T 1 2 GH 1 2 3 M 1 ABCDE 1 A 1 2 3 MN 1 ABCD 1 2 3 MN 1 2 3 MNO 1 2 3 4 T 1 2 3 M 1 2 3 4 TUVY 1 2 3 MNOP 1 A

"It is important to keep eye contact with me and not my PA."

Jeppe's PA explained that Jeppe would begin to sweat if the conversation was too fast or too intense. He needed it to be more chilled to avoid becoming exhausted.

Jacob was concerned: "How is it now?"

1 2 GH 1 A 1 2 3 MNOPR 1 ABCD 1 2 GH 1 2 GHI 1 2 G 1 2 GH 1 ABCDE 1 2 3 MN 1 ABCDE 1 2 3 MNOPR 1 2 G 1 2 3 4 TUVY

"Hard. High energy."

People do not always have time for his answer, so then he just answers minimally, without any personality and in a manner that he feels means he cannot express himself.

Jacob: "Is it annoying to have someone else finish your sentences?"

1 2 GHI 1 2 3 4 T 1 2 3 MNOPRS 1 2 3 MNO 1 2 GHIJK 1 2 3 4 TUV[W] 1 2 GH 1 ABCDE 1 2 3 MN 1 2 3 M 1 2 3 4 TUVY 1 2 3 MNOP 1 A 1 ABCDEF 1 2 GHI 1 2 3 MN 1 2 GHI 1 2 3 MNOPRS 1 2 3 GH 1 ABCDE 1 2 3 MNOPRS 1 2 3 MNOPRS 1 2 3 MNO 1 2 3 M 1 ABCDE 1 2 3 MNOPRS 1 ABCDE 1 2 3 MN 1 2 3 4 T 1 ABCDE 1 2 3 MN 1 ABC 1 ABCDE 1 2 3 MNOPRS 1 2 3 GHI 1 2 3 4 TU 1 2 3 MNOPRS 1 ABCDE 1 2 GHIJKL 1 ABCDE 1 2 3 MNOPRS 1 2 3 MNOPRS 1 ABCDE 1 2 3 MN 1 ABCDE 1 2 3 MNOPR 1 2 G 1 2 3 4 TUVY 1 A 1 2 3 MN 1 ABCD 1 AB 1 ABCDE 1 2 3 M 1 2 3 MNO 1 2 3 MNOPR 1 ABCDE 1 2 3 MNOP 1 2 3 MNOPR 1 ABCDE 1 2 3 MNOPRS 1 ABCDE 1 2 3 MN 1 2 3 4 T 1 2 GHI 1 2 3 MN 1 2 3 MNOPRS 1 2 3 4 T 1 ABCDE 1 A 1 ABCD

"It's OK when my PA finishes some sentences. I use less energy and [can] be more present instead."

But, his PA added, he has to stick with the right words since they are Jeppe's personality, and if not right, he pulls him up.

Martin and Jacob discussed how some of their friends talk quite fast and it can be difficult to butt in.

1 2 3 4 TUVY 1 ABCDE 1 2 3 MNOPRS 1 2 3 4 T 1 2 3 MNOPR 1 2 3 4 TU 1 ABCDE 1 2 GHI 1 A 1 2 3 M 1 A 1 2 3 MNOPR 1 ABCDE 1 A 1 2 GHIJKL 1 2 GHIJKL 1 2 3 4 TUVY 1 2 G 1 2 3 MNO 1 2 3 MNO 1 ABCD 1 2 3 MNOP 1 A 1 2 3 MNOPRS 1 2 3 MNOPRS 1 2 GHI 1 2 3 TUV 1 ABCDE 1 2 GHIJKL 1 2 GHI . . .

"Yes, true, I am a really good passive listener."

His PA predicted the last word.

1 2 GHI 1 ABC 1 2 3 MNO 1 2 3 MN 1 2 3 4 T 1 2 3 MNOPR 1 2 GHI 1 AB 1 2 3 4 TU 1 2 3 4 T 1 ABCDE 1 2 3 4 TUV[W] 1 2 GH 1 ABCDE 1 2 3 MN 1 2 3 MNOPRS 1 2 3 MNO 1 2 3 M 1 ABCDE 1 2 3 4 T 1 2 GH 1 2 GHI 1 2 3 MN 1 2 G 1 2 GHI 1 2 3 M 1 2 3 MNOP 1 2 3 MNO 1 2 3 MNOPR 1 2 3 4 T 1 A 1 2 3 MN 1 2 3 4 T

"I contribute when [there is] something important."

His PA said how rarely Jeppe was annoyed. He had been annoyed years earlier before realizing there was just no point. Jacob mentioned that sometimes he had something to say and had prepared his contribution, but then his friends have moved on and he never got to have his say. How does Jeppe experience himself when others have their own perspective and view?

> 1 2 GHI 1 ABCD 1 2 3 MNO 1 2 GHI 1 2 3 4 T 1 2 3 M 1 2 3 4 TUVY 1 2 3 4
> TUV[W] 1 A 1 2 3 4 TUVY 1 2 GHI 1 2 3 4 T 1 2 GH 1 2 GHI 1 2 3 MN 1 2
> GHIJK 1 2 GHIJKL 1 ABCDE 1 2 3 MNOPRS 1 2 3 MNORFS 1 2 GH 1 2 3
> MNO 1 2 3 4 TUV[W] 1 2 3 MNO 1 2 3 4 T 1 2 GH 1 ABCDE 1 2 3 MNOPR
> 1 2 3 MNOPRS 1 2 3 MNOPRS 1 ABCDE 1 ABCDE 1 2 3 M 1 ABCDE

"I do it my way, I think less [about] how others see me."

> 1 2 GHI 1 2 3 4 T 1 2 GHI 1 2 3 MNOPRS 1 ABCDE 1 2 3 MN 1 2 3 MNO 1 2
> 3 4 TU 1 2 G 1 2 GH 1 2 3 4 T 1 2 3 MNO 1 AB 1 ABCDE 1 2 3 4 T 1 2 3
> MNOPR 1 2 3 4 TU 1 ABCDE 1 2 3 4 T 1 2 3 MNO 1 2 3 M 1 2 3 4 TUVY 1
> 2 3 MNOPRS 1 ABCDE 1 2 3 GHIJKL 1 ABCDEF

"It is enough to be true to myself."

> 1 A 1 2 3 MN 1 2 3 4 TUVY 1 2 3 4 T 1 2 GH 1 2 3 GHI 1 2 3 MN 1 2 G 1 ABCDE
> 1 2 3 GHIJKL 1 2 3 MNOPRS 1 ABCDE 1 2 3 4 TUV[W] 1 2 3 MNO 1 2 3 4
> TU 1 2 GHIJKL 1 ABCD 1 AB 1 ABCDE 1 2 3 4 TUV[W] 1 ABCDE . . .

"Anything [else] would be weird" [with his PA finishing the word].

Even at the office where he worked, there were times when they saw the chair and not him. If he tried complicated sentences and thoughts, he could be misunderstood, underestimated, and feel negated.

The short interview was tiring for everyone, especially of course for Jeppe and his PA. During one break Jacob got up to watch Jeppe use the eye tracking system. As he went back to his chair, Jacob mutters, "It's fucking hard."

Later Martin and I sat together in his van as his PA drove us back toward the city. He was exhausted. He explained that he had been largely at home for the last two years because of the pandemic and was not used to meeting so many people. Neither he nor Jacob had created and conducted interviews in front of an audience before, and they had found it more tiring than they had expected, both cognitively and physically. Watching Martin up close made me realize how difficult it must be to talk, with so much effort, strain, and co-contraction of muscles. The spasms of the laryngeal muscles and the throat that then spread to the neck and body meant that by the end of the day his throat ached and his neck hurt.

I said how extraordinary it had been to see the four in the morning and then Jeppe in the afternoon, and how grateful I was to him and Jacob for allowing me a glimpse of their world—a world where talk was slow and effortful and where the idea of just chatting or gossiping hardly existed. He replied, "It is how we live. Every day. No one asked before. It is what it is. There is little reserve. Tiring. Vulnerable." I muttered that those with impairments can be like elite athletes, working at or near their peak every day just to get by. For once, he just nodded in agreement and rested in the seat.

11 Acquiesce with Silence: Cerebral Palsy

At a workshop on philosophy and neurology, they were difficult to miss. Minae and Michey live with cerebral palsy (CP) and are both philosophers and partners. Michey's CP—if he can be said to own it—is the commonest form, hemiplegic CP, affecting one side of his body, leading to weakness and lack of control down his left side. For Michey,

> There is no such thing as just going for a walk. As soon as I leave the house, I have to alter my position and posture, making tiny alterations every second. Every movement of my foot, swing of the stick, and grip on the stick handle has to be carefully coordinated. With no fluidity to walking I need constant planning of every move. I am always aware of my body as an object to be controlled, not really "part of me," and yet—and at the same time—it is certainly "my body" which I need to constantly reorganize. Other people appear as obstacles to be avoided, not just because I fear bumping into them and hurting myself and them. Even a hand offering help with shopping bags can appear hostile as it is an unexpected disruption to my "walking plan."

In addition to his very visible problem with controlling one side, he also has the unseen problems of reduced sensation and awareness of where his left-sided limbs are, making movement even more difficult.

Minae's cerebral palsy, cranio-oro-pharyngeal, is visible up to a point. Her problems are in controlling movements of the head and neck, and the muscles involved in swallowing and especially speech. Spasms appear first in the lips, tongue, and jaw, and quickly spread to other parts of the face, head, and neck and then to the whole body. Pain from contractions of her neck muscles over which she has no control adds to the irregular and jerky movements, which block speech further. Often tremor appears, and then her body freezes up. This begins if she thinks her speech is unintelligible and unacceptable to others, or as she tries to regain control of her speech.

These movements and the fear of not being understood have at times led her into being mute:

> There is no such a thing as just giving a speech. When I give a talk at a workshop or a conference I always feel uneasy because I not only feel nervous like everyone else, but also I know I am going to be frustrated at myself. Of course it might be because English is a second language to me but it is not only that. When I speak I have to think of which words I can pronounce easily. Even if I want to say more complex terms and long sentences, I cannot. If I think it would be impossible to pronounce, my body shuts down or becomes shaky.

We normally take for granted that words appear miraculously as we construct sentences without effort and often without attention, unless pulled up by the "tip of the tongue" phenomenon, when a word does not come to mind and interrupts the smooth flow. Rarely, unless one is an actor or mimic, priest or politician, do we consider *how* we speak; it just happens when we want. Minae has to think how she speaks and what words her body will allow her to say. Some words are too difficult: any word that starts with "s," "th," "h," or "ph," for instance. She thinks about the words each time before starting a sentence, and if they are too difficult, even in her own mind, her body shuts down or becomes shaky. This not only prevents speech, it spreads throughout her:

> When I relax I can speak. I know myself. But when I am tense, even though I am trying to speak I just cannot . . . which makes my entire body even more tense. When I am tense I get panicked so my muscle gets more tense and then I stammer and with that I feel pain in my neck and mouth.

She will shorten her speech and so reduce her thoughts and ideas. Long sentences are impossible, so long thoughts disappear before they become words. Some thoughts she keeps thinking and thinking, but then when she tries to express them, she needs words she can easily pronounce, and by then the thoughts have gone somewhere else and she forgets them.

Her problems are not just between her and the cerebral palsy; external influences affect her too:

> When I hear noise during my speech, it disrupts my whole self: I cannot continue speaking at all since my body shuts down. It is as if my brain just ordered me to stop speaking and to accept the forced silence, because my brain assumes that no one can hear my voice over the noise. I feel uneasy with people who feel uncomfortable asking me to repeat what I said and being with those who simply cannot understand me. Moreover, when I speak, my body moves involuntarily,

jerking back and forth, from left to right. Some people have assumed that I might be drunk with my unintelligible speech and involuntary movements. Without a signifier—a wheelchair, a walking stick, or a guide dog—I have nothing to indicate my impairment.

Speech is articulation and communication, but it is also performance and social. Being in front of people, exposed, is difficult for Minae. She cannot hide her feelings; it is as though the CP senses them and reveals all. Sensing any anxiety, it then heightens her dyscontrol, spasms, and pain. These arise, in part, from the fear of not being understood. Fear builds from the response of her listeners, and so she tries to avoid unintelligibility, often substituting words that are easier to pronounce for ones she fears will be too difficult to say. Sometimes she gives up and avoids talking altogether. She tries to cover up her condition, for instance, with one of her hands held just in front of her mouth while talking or writing notes. Recently, she has taken to wearing a surgical face mask in public the whole time regardless of COVID.

In her work, at least there is a more formal structure. Before she gives a paper she explains her disability and her vocal condition to her audience and asks them to read the subtitles of her speech on the computer screen in PowerPoint, which remains unusual in philosophy. Her audience often pays attention to the screen and the conference room is usually quiet.

Despite her extraordinary success in lecturing at all, she is enormously self-critical:

> I don't manage [speaking in public] at all. In fact, I always feel shameful to be in the lecture room. I use PowerPoint and some audiovisual materials. Before, when part-time, I used "text-to-speech" software. I wrote the whole lecture and set it to give a computerized speech. The weak point was this put most students to sleep! Now with so many classes to teach, I cannot prepare to that level and so I use my own voice. But I get so tired. . . . I wonder if I should keep on going.

She does not remember when she first talked and only realized her speech was abnormal when people asked her to repeat something. Her unaffected sister is a year younger but was treated as older; after all, she walked and talked first. Minae went to a speech therapist before primary school, but they focused on her breathing:

> I always wanted to be the "normal" but knew I could not be. People often stare at me and their gaze is embedded within myself.
>
> My CP doesn't affect so much my daily life. I can cook, with my helper who comes once a week doing the chopping and cleaning a bit. I can go to toilet by

myself; I shower myself. It is mainly the speech abnormality which limits my communication and ability to socialize.

Minae is Japanese and speaks and writes English as her second language, difficult enough even without her impairment:

> In my own language, people have the ears to listen to my voice and to try to figure out my speech. So I do not have a big problem to make people understand. However, in the train or bus, or in the public spaces, I cannot speak to them at all. When I hear some noise my body freezes. In addition, I have a neck pain when I speak so this stops me.

The relation between pronounced speech and a language's written form is complex. Vygotsky reflected that children learn to use spoken language way before they can write. One appears to come early and relatively easily, while the other depends on laborious periods of effort and repetition: "to learn to write a child must disengage from speech and replace the spoken word with imagery of words. . . . In speaking he is not conscious of sounds and quite unconscious of the mental operations needed."

Minae had singular problems in making this co-registration between spoken and written word. She could not write the Japanese characters because they were too complex for her to form with her hands, so it took a long time to acquire writing as well as speech. She is very grateful to her mother for introducing her to the English alphabet, which is far simpler to write. When she acquired the Japanese spoken language, she was with her mother and sister in a relaxed environment. She realized her voice was strange in comparison to theirs but did not feel uncomfortable or shamed. Outside the family, the problems were present from early on.

In the first year of primary school, her teacher asked students to read aloud one paragraph at a time in class. When Minae's turn came, she was so nervous she couldn't speak, even though she had acquired written Japanese. In contrast, before she learned spoken English she had learned the written language. Unlike Vygotsky's children, for her speaking was far more difficult than writing and reading. It remains problematic to read out loud in either language, with a noticeable time difference between thinking (reading the text visually) and reading the text aloud.

Having struggled to speak both Japanese and English, two very different tongues, in her own way, she has not always been helped by professionals. As a teenager she spent some time in Australia and volunteered to be a "guinea pig" for fourth-year speech and language pathology students. Immediately, she felt uneasy. They wanted her to pronounce each word in the Received Australian Phonological Standard manner, which was completely beyond her. Worse, when the students listened to the vowels and consonants she could produce and how she managed, she found it even more very difficult. They asked her to pronounce words that start with "p," "b," "t," "d," "k," and "g," and found problems with how she differentiated "p" and "b." They wanted to "fix" her pronunciation.

> I found it very difficult because my first language is Japanese and I have little knowledge of English phonology. They said that "p" was a voiceless stop and that "pat" and "bat" sound different. I thought I had pronounced "p" and "b" differently even when, obviously, they could not distinguish them when I spoke.

She had learned to speak her way, or at least the way she could impose on her CP.

Communication remains a huge problem for Minae. She moves her facial muscles into the right position to enable her to speak. She understands normal pronunciations of English and Japanese, but equally knows she cannot pronounce them as others do. It was impossible for her to say those words as the students advised because she has so many problems moving her mouth to form words correctly. Her problems with processing sound make things even more difficult. When she hears background noise, she cannot focus on listening to people speaking and cannot speak to them with ease.

Trying to improve the movements of her mouth and tongue, which some have suggested, is not helpful either since she has developed her own way of communicating and her facial muscles have adapted to certain ways of use. When the speech and language students asked her to say words, to close her mouth, or to stick out her tongue, she found it difficult since her whole body froze because she was nervous. They focused so much on her pronunciation that she found it difficult to talk at all and would speak English worse when with them. The students had tried, understandably enough, to help her by using the techniques they had been taught. But she had learned her own way, and they would have been better advised to work with her and learn from her.[1] While speaking Japanese is marginally easier for her, she feels more relaxed writing in English. Her problems with pronunciation may be similar, but she finds them a little easier in her native tongue. In Japan, people are native Japanese speakers and so more relaxed listening to her as well. In English, not many people understand and they have to attune to her pronunciation. Fortunately, Michey understands whatever she speaks, so Minae's body is relaxed and allows her to talk.

Her inner speech is in both Japanese and English, though she prefers to say she is dual-lingual rather than bilingual and she cannot mix the two. Inner speech comes first, and then she tries to find the words she can pronounce. If she cannot, then her body starts tensing:

> Obviously, my native tongue is Japanese so I have a larger Japanese vocabulary than English. It is easier for me to speak Japanese in this sense, since with more words it is easier to translate inner speech into spoken sentences. I use gestures and facial expressions, which come more naturally.

Speaking with people is not simply about the language used, or even its timbre, emphasis, and prosody. There is also an element of social performance, which in turn differs depending on whom you talk with and the context. Minae is well aware of all these factors. At home with Michey, she is relaxed and chats with little attention to the words, thoughts, and feelings that flow harmoniously. In contrast, in public, whether with students or even just asking the way in a street, others heighten her anxiety and can bring on the tension and freezing.

Unfortunately, as often is the case with cerebral palsy, attention toward to the body part that is misbehaving often makes things worse. There is a knife edge between no control and over-control: too little attention and the body wanders off on its own; too much and it tenses up in spasm. "I often have a problem between two extremes: no control and too much control, between too relaxed and too tense." It's not only in speech either; just contemplating what to say can bring it on. "I know I cannot pronounce certain words, but when I think of the word to pronounce my whole body freezes up."

Once in London she had prepared and wrote on a small piece of paper; "I have a physical disability, in particular a speech impairment, and cannot speak to you. Could you please tell me where Kings Cross Station is?" It was a small lie since she could speak but she felt uneasy speaking in a public space and anyway assumed no one would understand her. It was easier to be mute. She also knew that her whole body would freeze up if her speech was rejected by the listener. Despite this, most people just ignored her before finally someone helped. She wondered if it would have been easier if she was recognizably disabled; then people might have treated her better and not been shocked. She does not want to scare people.

Minae is helped hugely by any social setting where she feels relaxed and comfortable. However, in reality, such times are rare. What she has found helpful is computer technology and the internet. So, to help her communicate, it is less important to fix her speech than to help with alternative technological devices, a quiet space, and other methods of making her feel relaxed. She finds email and online chat systems wonderful. Communication is no longer ambiguous, and with that her frustration is reduced. With email her CP is no longer relevant or in charge.

Sometimes she cannot identify certain words until she has positioned herself in the relevant English or Japanese linguistic space; other times she cannot pronounce certain words so tries to think of other, easier ones to pronounce. When she is tense and cannot speak, in her mind she goes, "Ahhhh . . . this happens again! Why can't I speak?" She feels isolated, with an empty space inside her as she tries to connect thought and language:

> In this empty space emotion comes first, unformed in language. Do I make sense of it? I love talking and expressing, but it is so time-consuming to explain my thoughts and find the quiet place, the place where I can be relaxed and express [myself].

As we have discussed, much thought occurs before and without language. But to communicate, language is needed, and the emergence of words from thoughts, a miraculous materializing, remains a hurdle for Minae much of the time. She describes herself as having two opposites, not in the sense of being bipolar but in her communications with others:

> It is like having two different selves . . . one who acquiesces with silence and another who wants to talk more . . . opposites within me, passive me and active me. My students talk to me and most lecturers are talkative, but I am not . . . so they need to search for their own answers, rather than I offer them an answer. Just to hang out and chat is always an effort, always a challenge. Sometimes I feel so sad at how difficult it is; there are few good sides of being silent.

She is very aware that speech has prosody, pauses, volume, pitch, pronunciation, and the 1,001 nuances, gestures, and expressions that accompany it and that allow us to show mood, feelings, and other inner thoughts and that, unlike the written word, this all happens in real time. Perhaps surprisingly, she sees some of these in writing too:

I see pauses and pitch in written languages. As I read your messages I imagine your gentle voice. My problem is that I cannot control the muscles that allow me to talk and sing. But I also realize that when I can talk without any tenseness, people around me are also relaxed and listen closely. That is very special. There is then a kind of aura I feel. This aura or atmosphere is very social.

Many people, including philosophers, authors, cognitive scientists, and doctors, have debated and questioned the relations between language and thought, from William James to Tolstoy, Vygotsky, Merleau-Ponty, and, more recently, Steven Pinker and Oliver Sacks. While one view is that thought is within a person and language a readout of some aspects of this, such a directional view of the two seems overly simplistic. Merleau-Ponty suggested that in some situations speech could articulate and even lead our thoughts, as in "We often don't know what we think until we speak it."[2] Whatever the relation between the two, speech is also as much about gossip as grammar, about social interactions and getting on. But Merleau-Ponty's hunch does reflect Minae's biggest fear, that people may assume that without speech she is thoughtless, unintelligible and therefore unintelligent, mute and therefore dumb:

Speech matters to me. It is a kind of performativity. As a result of my difficulties, people often treat me as a someone with intellectual disability. I do not like meeting strangers the first time.

At one level she is very aware that "normal" women speak in certain ways that she cannot, so her voice—and she herself—is sometimes rejected and ignored. Above that she has a deep frustration that though her thoughts are not constrained by language, their communication is:

I avoid words I cannot pronounce. I make my speech short and so shorten my thoughts and ideas. My thoughts and arguments are constrained not by knowledge or my creativity, but by the link between me and the vocalization part of my brain.

The spoken word or, more correctly, spoken sentences are about thought and language and conversation and turn-taking, about people learning about each other and talking in a relaxed social to and fro. Minae finds all talk difficult, but social talk that for most people comes naturally and with ease is no easier than other forms. Remember, she has to manage her unruly head and neck musculature, with some but not too much attention, and manage her difficulties with articulation together with the tension and sometimes pain, while considering all the time which words she thinks she may be able to say, knowing that even thinking of difficult words can upset the balance. All this as she listens to other people's chatter, hoping that the ambient noises don't intrude on her comprehension or add to her anxiety. Like a juggler on a tightrope, she also has to make it all look relaxed to blend in as best she can and to ready herself to reply as nonchalantly as possible to avoid disrupting the atmosphere.

I spent time with Michey and Minae in Japan. We ate out several times; they showed me around a Shogun's castle and delivered me to the university in the suburbs of Osaka for my lecture. It was summer, sunny and hot. As we walked, Minae was always five meters ahead. What fluidity she lacked in talking and eating or in gesture was found within her walking. Graceful and lithe like a runner, she seemed to float through the world so light on her feet she would leave no trace on snow. I walked behind with Michey. His fluency was in talking and discussing; he was, after all, a philosopher. In contrast, walking was all effort, concentration and sweat. While Minae was ahead and away, Michey lumbered as though under a different force of gravity.

None of this mattered. Together they were a whole and complete. Observing their difficulties with walking or with eating or conversation, seeing how much effort they put into what we all take for granted was a strange privilege and education. Our family takes its holidays in West Wales, where Atlantic waves crash, the weather is always mixed, the farming hard, and wind-sculpted trees squat close to the ground. There is something about places where the living ain't easy, and a strength and quiet dignity to those who work hard just to get by. Minae and Michey expend effort to do what we take for granted: to walk, dress, eat, and talk. They relish life all the more for it never being facile, and they too have a dignity and integrity that is moving to observe.

12 Walks on the Wild Side: Nonfluent Aphasia

One of the most common causes of impairments following a stroke involves the areas of the brain responsible for the production and/or interpretation of speech, gathered together under the term "aphasia."[1] Despite widespread use of drugs to reduce blood pressure and to mitigate the effects of heart arrythmias, strokes are still a major cause of death and chronic impairment. In the United States it is estimated that around 750,000–800,000 people have a stroke each year. "Stroke survivors" number about five to seven million, of whom 2.3 million are significantly disabled.[2] Language areas in the brain are widespread, reflecting the importance of language and speech production and reception. Impairments in speech occur after strokes involving mainly the left hemisphere of right-handed people, in the temporal, parietal, and frontal lobes.[3] It has been estimated that following all strokes between 25 and 40 percent of people have a form of aphasia, meaning there are maybe just under two million people affected in the United States alone.[4]

Aphasia is not a singular condition, but rather reflects the cortical damage after stroke to the various areas and to the connections involved in language production and comprehension.[5] Because of the individuality of strokes and a person's response, Howard and Hatfield suggested there might be 16,383 varieties.[6] Less dauntingly, there is good agreement that there are certain categories of aphasia.[7] These are divided into two main types: those in which speech is hesitant, nonfluent, and effortful, leading to its other name, expressive aphasia, and those in which speech appears fluent and in which sentence structure appears relatively preserved, though, extraordinarily, meaning within the words and sentences is often lacking.

The nonfluent aphasias are further divided into those in which language understanding or reception seems largely intact and those in which it is not. Within the latter group global aphasia is perhaps the most severe, with serious expressive and receptive language impairment and communication possible only through facial expression, gesture, and voice intonation. The severity of this condition suggests the stroke has affected all parts of the language brain, but also that other aspects of speech—prosody and its accompanying facial expressions—are located elsewhere in the brain. Those with language understanding remaining are subdivided further into those who find repetition of words difficult (Broca's aphasia) and those who can repeat (transcortical motor aphasia). One of the important aspects of repetition is that it suggests a short-term memory for speech production, which is essential to build up a full sentence or paragraph and hence extended thought or argument. Recall Jeppe with cerebral palsy and his PA. If neither could remember long strings of letters and words, then they would not have been able to communicate.

There are various levels of severity of all these types, hence the more than 16,000 varieties. People with Broca's aphasia usually retain the use of the main content words, nouns, and verbs but lose other aspects, such as prepositions that place a word, for example, *on* pain *of* death, and some conjunctions, for example, "but" and "if" (but not "and"). They also mis-say words and sentences. This can make their speech abrupt as well as effortful and sparse, leading some to suggest it has a telegrammatic aspect. People with Broca's aphasia are often able to understand others' speech as long as the grammatical structure of a sentence is relatively simple. They are also aware of their own impoverished speech, leading at times to frustration and depression.

Fluent aphasia is divided into those cases where people can largely understand language and those who cannot. Within the former is conduction aphasia, in which people have difficulties in word finding and in repetition, and those with anomic aphasia, in whom repetition of words and phrases may be preserved but who have word-finding difficulties so will use fillers and circumlocutions instead.

The last group are those fluent aphasias, sensory aphasias, in which understanding of language is profoundly impaired. In the most well-known of these, Wernicke's aphasia, spoken language is not understood, whether that of others or the person affected. As a consequence, because speech production is unaffected but its self-monitoring is affected, people with Wernicke's aphasia will speak fluently but with profoundly impaired content. They may produce paraphasias, substituting one sound for another, for example, box with cox; replace one word for another, for example, blue with green; or even produce entirely new, nonsensical words. Their speech may be fluent and have prosodic content but with disordered grammatical structure and without meaning or semantic content. Without self-monitoring of its output, the speaking brain goes haywire. With an apparent lack of awareness of their own speech, those with Wernicke's may be less prone to frustration or depression than those with nonfluent aphasias, even though they inhabit a world with reduced understanding of others and without insight into their own expression. This lack of self-monitoring is also reflected in poor repetition of words and phrases, whereas in transcortical sensory aphasia it is relatively intact, though people merely echo words, phrases, and even questions rather than answering them.[8]

Last, we must consider speech apraxia. An apraxia is a disorder of the planning of complex movements. A person with dressing apraxia may be without weakness or loss of coordination, but no longer have intact the motor programs that enable them to dress, put an arm down a sleeve, or tie a shoelace. In speech apraxia, people know what they want to say, have the words in their heads, but cannot make the movement part of the brain articulate them even though otherwise the pharyngeal muscles work normally. Willem Levelt, a psycholinguist, developed a model of speech production in which there are several levels.[9] The first conceptualization involves the development of what is to be expressed; the second formulation involves the selection of the words, grammatical structure, prosody, and other elements through which the concept is to be expressed; and the last articulation involves sending the appropriate motor commands to the speech apparatus. Speech apraxia is thought to be a deficit at the articulation level. Speech becomes effortful with hesitancy, inconsistency, and attempts to correct errors.

The relative frequency of the various types of aphasia has been estimated as being global 50 percent, motor/expressive 28 percent, sensory receptive 11 percent, and others 8 percent. Modern neurolinguistics has allowed more detailed understanding of the types of aphasia and their deficits and thus allowed advances in understanding how the brain organizes speech production and reception. This short account cannot do more than give a sketch of such a field.

This medical and epidemiological view is one way to look at stroke, but another is to explore the experience of the condition, and then one almost needs the inverse of statistics: the personal accounts of aphasia and its consequences. Obtaining narratives from any person with problems with speech has its difficulties, and these are perhaps heightened in people with aphasia. These accounts, however incomplete, allow glimpses into the consequences of living without ease of speech following stroke, built up from the people so affected and their loved ones. Having interviewed a number of people, mainly couples, the following accounts explore both nonfluent and fluent aphasias. Stroke is a severe singular catastrophic event. If a person survives, then it is rare that the resulting deficit is permanent and unchanging; improvements both small and large can occur. So the nonfluent group is divided between those who have recovered to some extent and can reflect on their acute episodes, and those who remain stranded within their communication problem.

Walk on the Wild Side

When Steve, in his late forties, came into the kitchen one Saturday feeling sick and feverish, his wife Natalie presumed he had caught something and sent him to bed. Six hours later he was still asleep and she was worried. She woke him to find that he had problems seeing and talked nonsense. As a hospital nurse she knew what it was. While Natalie called an ambulance, Steve went to the loo and then dressed:

> I'd been ill all day and was woken by my wife late Saturday and she called the ambulance and the um, the, err, the what you call the men, paramedics, said they'd get me into hospital. I dressed myself, walked myself out to ambulance and sat there in the ambulance and then walked into and that's all I remember.

He actually walked into the side of the ambulance, which he could no longer see. When he tried to answer the men, Natalie saw how he talked to the wrong one. He had developed a visual field deficit so that he could not see to the right. Scans showed an intracranial bleed, so he was not a candidate for clot-busting drugs:

> Even today I have lost my vision on the right-hand side. I can only see your right eye, I cannot see your left eye, that's gone.[10] I couldn't walk at all to start with. Natalie's a nurse and goes to the stroke unit anyway so she helped me during the day. I could not talk, I could not read. I had lost the ability to write. Strangely, I could hear everything on radio. I was struggling . . . could think straight and

could compose a sentence in my mind but could not say it. Trying to get my head straight now . . .

Ten days in hospital but when I woke up first morning in in in the normal out there at the main ward I, I, I could not walk properly, but I was not going to let myself do nothing, so the first morning I climbed out of bed, got my T-shirt and trousers and I took my washing kit and got the bathroom and washed.

Walking was difficult because his right side was weak. But well-motivated from the beginning, he forced himself out of bed. By three months he was walking. Once home he started walking thirty meters and now he's up to eight or nine miles each day. With the lost sight he can no longer drive, so he walks.

He could not remember names, even his wife's. Formerly in the army, he did not know his forces number. Natalie remembers how scared he was. Initially, he thought he had died and was in another's body[11] and was angry because he thought no one was doing anything. All this was nine months ago. His weakness in the right leg recovered, but speech has been sluggish and he still struggles with names. As Natalie remembers, "It was several months before he had the words to explain what he was experiencing. That must be so difficult." Speech still goes "all weird" when he is tired. He also prepares what he wants to say but then it disappears:

I knew you was coming I prepared. . . . I lose that and don't talk as much. One it can be difficult to keep up if they talk too fast and also because I lose what I want to say halfway through and I have to stop because I forget what I want to say . . . well I cannot find the words, so then I stop. It makes me feel a fool. So I don't bother talking. I don't finish off a conversation or the details.

It was worse, but I still have these problems with finding words, what I had for breakfast, sometimes I'll have porridge. It's come out this time but tomorrow I may say I had. . . . [flicks fingers . . .] it comes and goes. Someone finds a word and throws it out the window. I see a bowl of gray stuff but I cannot find the word for it. I have found it by waiting, helping by knowing, visualize what I had this helps. It is random it goes it is gone. It takes a while to say. I may say its porridge to you and in five minutes it's gone. Once I've got the word there is no problem saying it.

If I got on a bus I could not say where I wanted to go. . . . I would not have the words . . . it's that place down the road. Very frustrating. I'll find it again. A few weeks ago I wanted a battery for watch. Can't remember the town . . . it's gone. It was not till I walked to the shops as I walked past, it had a sign, I knew where I wanted to go but the name was gone disappeared.

Equally problematic is reading. He could not read anything when he came out of the hospital and still struggles both with the act of reading given his sight and with recognition. The hemianopia means he can see one side of the word or phrase and has to remember that as he moves across to read the other and then put the two together, if he can:

> I can slowly read an email. Every word I look at, it's it's like a new word. I have to look at it again and work it out. . . . I have to start from the beginning each time. The aphasia damages the reading side, my memory and recall are a problem. If I've managed a line I look again and forget. When I go to read something, when I look all I see one side and when I've got that the other part goes. I cannot see both sides together and then I have to go back to read, to read and then remember the one part, it takes a few minutes and then I can spell that. But the next one I cannot remember cannot follow one sentence to the next.

His spelling has improved. He can now spell "aphasia" and uses the internet and Alexa. But it remains a trial, as though all the words have been thrown on the floor and jumbled up and mixed. When writing he may forget one letter:

> Say an "H," at the start and see that "a, a, a a," if you don't get the "H" the word's wrong. I'll say a word and say it and it doesn't sound right, then I'll realize it's a "P" not a "B" I'm starting to spell it correctly. It's all part of the thing I just lose. Some words I can't find, some longer words, some I only see half of it to start off with and the English language is funny how you say it and certain letters are there not pronounced.

I show him a long word: "tomorrow." Initially, he mistakes the "t" for an "l." "Lom . . ." I suggest he looks at the first letter again. "T not a L. Tomato? Tomato?" I give the context, "He could be in hospital for a week, unless the carers can start . . . tom . . . ?" "Tomorrow. That's a good example. If I go 'tom' first, then I lose 'morrow' and don't see it."

He also has problems in recognizing objects and finds complexity baffling:

> If Natalie wants something from the fridge I cannot find it. "Get me an onion." As I'm going, I know the word but which is the onion? Sometimes I don't recognize an onion. First time we are shopping a minefield so many things in a shop. So many things.[12]

Being able to work again is a big worry with many factors involved. "I need my vision, I need my recall and I need to read and write even as a mechanic. I'd be told to do a gear box tomorrow and I would forget. I would not remember how to do it either." Recently, he had replaced a belt on a lawnmower. "I got the lid off and then got a new belt, I had to think where it went. It took twenty minutes to work out what to do and then it clicked

and all was okay. Put it back together again. It should have taken ten minutes now took an hour."

As a nurse, Natalie combines insight, love, and support with the necessary steel:

> He repaired the lawnmower, which he had never done before. Okay, he could not
> find all the screws but that was his vision. When the occupational therapist came
> round they asked him to make a cup of tea. He could not tell them *how* to make
> it but he could make it. He can do it.
>
> He can feed himself and eat, do shoelaces. He moves okay, just a bit slower.
> He's not allowed to drive but he's moved the car on the drive[way]. I come back
> and it's moved. But if he worked, he'd be too tired. He could do a simpler mechan-
> ical job but not read a computer. He can pace himself at home without no pres-
> sure, but any job and any pressure he loses words. Even his physical posture alters.

She has seen a vast improvement. He can walk nine miles on his own
and he is never lost in familiar places. The problems remain in reading and
driving. The latter, she says,

> is massive. I'd be quite happy in the passenger seat but not him. We have gone
> from a three-car family to a one. He joined the army at seventeen and learned to
> drive a tank before a car. If you gave him a tank, I think he could take it apart. It
> would take a while but he'd be able to do it and fix something.
>
> His vision is a big problem; he used to be able to drive a tank onto the back of
> a lorry without any wriggle room looking through a small letter box [opening].
> He now does the housework and the ironing . . . and always has to remember to
> scan to the right . . . that is a dirty phrase in this house. Watching a TV program
> he watches the left half of each scene. I put everything to his right at tables and
> desk. Ironing is good because he has to iron both right and left sides. As a nurse
> my uniforms have never looked better. For cleaning I have to label which product
> is which. Otherwise, he would have no idea what to use on the mop. But I realize
> if he worked he'd be too tired.

Natalie did not realize before how important reading is. In a film or at
the beginning of a TV program she has to remember to say it. He cannot
even follow a shopping list. Sometimes she'll draw pictures instead but he
is bewildered by the number of things in a shop. "Asking him to get something from the fridge he cannot find anything or remember what I asked
for, but I am not going to stop. Doing it all for him will not help him. I feel
guilty that I can just enjoy reading a book and he cannot do that. I feel sad
that he cannot sit on a beach and he cannot sit and read. He's like a child."
She has found it exhausting, watching him relearning to recognize letters
and then spell letters:

Living with aphasia is tiring—for both of us. When he's looking for words it's best to prompt him rather than tell him. A lot of people want to jump in. His sister and mine cannot wait for him to find a word and so prompt him, but waiting is not a problem. I find it hard to do all his reading for him. I get home tired after work and he's there and needs a letter read. He needs to know what it says. He needs to know. . . . Like being a little boy again but with a knowledge of what's lost.

We both like museums. We went to the War Rooms. Fabulous but a lot to read. They have an audio guide and he could use that . . . but I still ended up reading loads and loads. Even going to get an ice cream I have to read the flavors out since he needs the choice. He deserves it. One shop had twenty different flavors. There used to be only three; vanilla, strawberry, and chocolate, or a Neapolitan [with all three together].

But she does admire his positive attitude. "He'd to say, 'If you've woken up, it's a good day.' He plows on. His first wife died of breast cancer and I am a palliative care nurse. You go on."

This is not to lessen the problems they both face. Steve, once more:

I have been told aphasia is incurable. But you can work round and educate it. So frustrating sometimes. You don't feel the same person. Confidence goes down. Natalie has to read everything. I hear her but then forget and cannot say things. . . . I get confused so easily. Going away from the physical side the mental really throws. I have broken down a few times. I feel guilt that Natalie does everything. She has moved from part-time to full time. I will lose my wages. Fi, feel guilty. She has to do everything, she has to write everything.

He finds one of the hardest things is not knowing how much recovery he might expect, and how much he'll find ways of working round:

It's a horrible situation, a horrible disease. I have to work hard to keep going. It is so tiring. If I go for a walk or do speech and language for an hour, I am shattered and I fall asleep. Can be depressing if you allow it. I say to Natalie, "It is what it is. I must not let it eat me. Do the best I can. Try to enjoy it. Frustration, can't drive a car nor ride a bike. They do get you down and then you think you are alive. And a lot of people don't get that far."

Natalie has managed to find some compensations:

If comfortable with people he is still a chatterbox. He says he is quieter . . . but he's not. Steve had never been the most self-reflective person and he now reflects far more than he ever did in the past. He'll leave things around. He does a man look. He'll say to me that this is the aphasia and I say, "No, no, you were like that before the stroke."

If anything, she thinks he is better with friends than before. Before, he would interrupt, which Natalie found "infuriating," whereas now he takes more time and thinks about things more. His condition has made him more responsive. He may struggle with more complex matters, but still does remarkably well considering how much he's lost. Natalie again:

> He's better one to one. Socially, a lot of my friends are nurses so he feels safe and they give him time. My mum was a special needs teacher so she is very patient, and my son teaches English as a foreign language so he is patient, and my daughter just joins in and they became a pair. Morphine, (for the headaches) turned him into a hippy. Everything was, "Hey man." In a loud voice, he'd say, "Hey man, take a walk on the wild side." My daughter had socks made with that on the bottom. A sense of humor helps. People sometimes seem quite shocked at our sense of humor and repartee; it is love, but tough love.

Aphasia, in his case Broca's, rarely comes alone. He has naming and word recollection difficulties and continues to have speech production problems, together with visual hemianopia and difficulties in reading and writing, and the cognitive sum of these mean that making sense of the world becomes more difficult:[13]

> I appreciate what you are trying to do, getting into my mind. I can't explain what my problem is—one of my biggest frustrations. I have no control over what my brain is doing. I have damage up here, which may be slowly recovering and no control and I feel I have no control over aphasia and throws memory out the window. Recollect, recollection out the window, not control. It's less what I am thinking but how I can express it to other people. I can always understand to myself. I know I want that is there even if not in words. I know I want to watch the football next Saturday; I can walk five miles to get there. All my memories are intact. I find a way round.

Compared with many who live with aphasia after stroke, he is comparatively young and retains insight into his condition and mental resourcefulness. Couple these with his wife and family, and despite the difficulties imposed by the stroke, he and Natalie will surely continue to find ways of making it work.

Nicer

Bill was a trainer in leadership for banks when suddenly he had right-sided weakness and everything became a blur.[14] The only thing he knew for sure initially was that he could no longer talk:

> I knew something was wrong, but was not sure what . . . I was very frightened, everything blurred, with a jumble between thought and speech.

His recovery in hospital was actually fast: he was transferred to a rehabilitation ward from a high dependency after two weeks, and two months
later he was home. He recovered walking, then arm function, and then,
sometime later, speech after a fashion.

He remembers how frustrating it was to neither talk nor understand
what people were saying. "My brain could not cope with words." "You want
to shout your anger and frustration, yet people were helping me so much
you couldn't get angry. The speech and language team (SALT) was amaz-
ingly good at little things, mainly talking with me. Words returned before
meaning."[15] Bill remembers knowing what a cat was, but not knowing what
the word "cat" meant. Every day the SALT team would come around to
teach him words. "It started to get better about five weeks later. I could
understand what people were saying and still could not talk. Watching
TV, I had no idea." Once home, friends brought lots of books, including a
children's alphabet. He remembers relearning words, "xylophone" being a
favorite. One SALT member visited out of hours after work, for weeks. As
Bill recovered, he felt as though bits in his speech brain would pop back,
"like a jigsaw puzzle."

Initially, he could express only one thing in one sentence and then pause
for the next sentence. He could only do chunks and his thinking was in
small chunks too, with the two not always connected. He might forget the
end of a sentence. "My friends laughed. I am talking and say I've forgotten
what I am talking about. You have to train yourself that it does not matter."
He is still frustrated by thoughts he cannot express but is becoming reconciled. "Even now [several years later] round six o'clock I am tired and my
energy goes down. But I'm happy because I have a good day, but I don't care.
Tomorrow I'll be great." Bill still finds it difficult to tie meaning to words:

> I simplify. You know what you go to talk something and then you can't remem-
> ber, so a word comes out and the word, I am not explaining, like people who have
> a stutter and they struggle for a word like it, and I would say, if people are . . . just
> said . . . Before the stroke, I would never, but now I don't know that word and
> another word takes over.
>
> I would say things I did not mean, the first weeks I would say "No" but mean
> "Yes." I did not know I was saying it wrong. The family was desperately trying to
> help. There is a gap and another word will take it over. I could not talk, jumbled,
> I remembered that and I felt more better. But it was it's amazing now you can
> [cover] for everything, at least you can do something but because you brain can-
> not cope with words, I could not . . . [always say what I wanted?]

I would make a "T" to ask for tea, even though I could talk you could get it wrong. People could pick up stuff. My friends realize I am good now, because of the training. Amazing how to ad adapt, otherwise you . . . Older people find it hard.

After my stroke my friends wondered if I would be the same person, and the first weeks and months I was not, but now they say I am back. That arrived. I've got I've got I've got quite a childish humor at times and even more now than before. I love stupid things like the "Carry On" films. I think, one thing I will say, I feel gladly now more for people. I do not suffer fools gladly now, pleased I don't have to do that now, as when I was a trainer.

Sometimes I would [open my] mouth but no voice, the thought but not the talk. Family and friends were quite shattered to watch like a hawk to see what I mean. . . . Now I realize how much you see in body and face; I am more aware of non-language. Basic expression, not through words. Facial expression hugely.

He still finds crowds hard and can follow only one conversation at a time. "My brain will hear little bits and its jumbled. Some people can have three conversations. I cannot." He now visits his old hospital to help people come to terms with their aphasia. "Sometimes they [nurses and relatives] will see someone [who] can't talk and point them at the TV. . . . They cannot talk, they are angry and they cannot understand TV either." Instead,

Sit down with them, talk with them slowly till they understand what is going on. If they cannot talk, show them, engage with them. "Coffee?" say "Coffee, coffee?" I was so lucky to have family and friends come in and talk to me even when I could not reply. Others should structure their words and thinking to their [loved one], to be more human.

Guy Wint, who became severely aphasic after a stroke, makes a similar point about the importance of reaching out and talking:

I could not speak; words tumbled over themselves on my mind, but I could not get them onto my lips. When they came I had sometimes forgotten what I wanted to say. This division between meaning and words, normally one, is one of the plagues of the stroke sufferer, with lack of vocabulary. The victim is bewildered because he cannot explain his predicament. Sometimes it is said the patient is anxious to be left alone. Nothing could be further from the truth.[16]

Wint was not unaware of the problems inherent with this: "Having a stroke is so very awkward for social intercourse. . . . The non-stricken must be sympathetic, and yet does not know what it should sympathize with."[17] Bill was particularly saddened by the way he had seen fellow patients become depressed and even divorced—so much so that he now works as a counselor. "There is a big incidence of depression and family breakup after stroke. My second group is about the family. I am really so sorry when I do my groups for the families; some are so angry and frustrated."

Depression is certainly increased in those after stroke and related to their incapacity and need for assistance, as shown in one study.[18] It is more difficult to determine whether someone with aphasia may be depressed because of their communication difficulties. There is, however, evidence from one study that around 20 percent of those with aphasia are depressed, with another 20 percent showing signs of subthreshold depression when assessed carefully.[19] These are high numbers and greater than in those with post-stroke depression without aphasia. Aphasia is a huge additional burden over other movement impairments.

As we chatted, Bill seemed at ease with his new way of living. He knew he could not express himself as before but was at peace being less frantic, with more simplicity and more time. His stroke had led him to a different place, which felt better. "Before I was very focused on everything in work, always work . . . now I'm more chilled. Good for me. I'm nicer."[20] His stroke has deprived the banking world of a trainer, but Bill is clear that he enjoys his new life more. Almost as an aside he mentioned, "Now I also have vasc[ular] dementia. Strange, some days it's the dementia more than the stroke."

A large stroke has been found to increase the chances of developing dementia around six- to seven-fold in the subsequent five-year period.[21] All types of stroke double the risk.[22] In one study, around 10 percent of people admitted to hospital with a stroke went on to develop dementia in the first year after,[23] while in another study dementia was found in a quarter of older patients after stroke.[24] The factors involved are unclear, though age, pre-stroke cognitive decline, and brain atrophy are risk factors. Some strokes are small and lead to multiple small deficits that in time lead to vascular dementia, and it is this type of dementia that is slightly more likely after a stroke leading to aphasia.[25] In other cases large strokes will affect global brain functioning. As with Steve's visual field loss and reading difficulties, post-stroke aphasia often does not arrive alone.

Steve has been exploring how to live following stroke for under a year, and Bill for two years. We now consider a woman in her forties who has lived with her stroke, which led initially to a dense nonfluent aphasia, for over half her life. She can remember when it happened, how she recovered, and how it continues to affect her each day.

13 Uncontrollable Dog: Nonfluent Aphasia

Faye's severe stroke arrived in her early twenties, and now, two decades later, she can look back at the experience and at its residual effects. She was twenty-one and an assistant manager in a bank when one morning she developed double vision. She thought it was a migraine, so her boss suggested she go home. Her sister worked opposite and agreed to pick her up. Faye then waited outside from eleven till five to be taken home and thought nothing of it. Logic was beginning to leave her.

Once home, her speech annoyed her partner. "For goodness' sake just talk properly." It was when her walking was affected that he took her immediately to their GP.[1] He dismissed it as a bad headache. They went home and she had a bath and went to bed, though by now with tunnel vision and pins and needles down one side. She still thought it was a severe migraine.

In the morning, she woke without speech. "Stupidly," her partner just got ready for work and remembers her waving goodbye from a window. Then thinking again and realizing something was seriously wrong, he phoned her sister and she and their mother went round.

Despite being unable to speak, Faye did not recognize this:

> I was thinking away in my own head albeit illogically. I got out of bed and waved goodbye to my partner. I remember everything, walking into bathroom wobbly, and going to the loo and then standing up and knocking all the pictures off the wall. There was no thought in my head that this was not right. I saw it all but did not question it.

There was a knock at the door and as she opened it she collapsed onto the floor, totally naked. Her sister dialed for an ambulance and when it arrived the paramedics presumed she'd taken drugs. Faye remembers being both embarrassed and angry. Not once, though, did she feel ill; even in the ambulance she remembers just thinking she'd never been in an ambulance with blue lights before. In the emergency department she was in one cubicle and her mum—who'd fainted—was in the next.

In intensive care it took some time to establish she'd had a widespread stroke affecting several areas, leading to eye gaze problems, pins and needles and paralysis down the left side, loss of speech and swallow, and difficulty with handwriting and calculation. No clear cause was established, though her use of oral contraceptives and her smoking were thought relevant. She remembers little of the next days. Doctors inserted a tube into her stomach through the abdomen. Without speech or swallow and with a dense paralysis down her left side, everyone thought she was in for the long haul.

The bewilderment of an acute stroke was captured by the playwright Arthur Kopit:

> The moment of a stroke, even a relatively minor one, and its immediate aftermath, are an experience in chaos. Nothing at all makes sense. Nothing except perhaps this overwhelming disorientation will be remembered by the victim. The stroke usually happens suddenly. It is a catastrophe . . . the "victim" cannot process. Her familiar world has been rearranged. The puzzle is in pieces. All at once, and with no time to prepare, she has been picked up and dropped into another realm.
>
> An explosion quite literally is occurring in her brain, or rather a series of explosions: the victim's mind, her sense of time and place, her sense of self, all are being shattered if not annihilated.[2]

Faye remembers little of the next weeks. A diary she started with her care team begins two weeks later and is full of reports from physiotherapists, mentioning how she transferred from bed to chair with help and using her left hand a little. With no words she used an alphabet chart, pointing to each letter. "It was so annoying because I could not spell. My family asked what I had been doing and I tapped out THISIO . . . THISIO (physio). They had no idea." On the reverse of the alphabet chart her sister put lots of swear words. Faye also had a bit of a crush on her neurologist so her sister put loads of stuff about him too.

Three weeks after her stroke she managed to tap out her Golden Goals:

> Move thumb upward
> Go to loo on own
> Talk (a lot)
> To be able to walk home
> To drink water, followed by lager
> To eat food, particularly pizza

To move my left side as well as my right
To be able to write
To able to brush the hair on the back of my head
To be able to kiss and cuddle with my partner and blow him kisses.

"To be able, to be able . . ." Soon she was transferred to a stroke unit and then to rehabilitation, each move closer to normality. The diary mentions her first bath, applying deodorant herself, washing herself, wheelchair skills, bum wiggling on the floor, then standing and moving her thumb. She was given a Lightwriter, paid for by the bank, which enabled her to tap out her words, which were then spoken in an American accent, which amused her. But what Faye remembers most is an older woman doing her hair and nails. It was her babbling away as she did them, talking to her, which helped Faye more than she'll even know. She was wheeled outside to see the birds. The ward staff loved her since unlike most of the other patients she was young and bright.

She also surprised them with her rate of recovery. After four weeks she was standing independently, while at five weeks she began handwriting and eating "mashed potatoes, carrots and sprouts. I even had a cup of tea." The diary details her tying shoelaces, doing her own hair, standing independently, writing, using her left hand—each advance hard-won with huge effort and concentration. After one session the diary entry from the physiotherapist read: "Afterward, Vicki, Faye and I had a little cry." She relearned to put her whole foot down, stabilize her hip, and not let the knee buckle. By concentrating on its individual components Faye relearned to walk. By six or seven weeks she spoke, clearly, though initially with a foreign accent.[3] "I sounded like Steffi Graf."

By eight weeks she rode a bike, "so dangerously, with a physio running behind me I nearly killed us both." By nine weeks she went shopping. Bus trips, walks out, and cooking followed. After ten weeks, once the percutaneous endoscopic gastrostomy tube, which had allowed them to introduce nutrients to her stomach when she was unable to eat, was out she was home, with a host of touching goodbye notes from nurses, occupational therapists, physiotherapists, and other patients: "remarkable determination," "admiration for your courage," "inspirational."

A couple of months later she returned to the bank, though she was only allowed to do the more mundane jobs. She opened the post each morning, which she had always liked. For some reason when she opened a complaint to the bank she decided to reply herself, handwritten with appalling spelling. The person complained again and it cost the bank a lot of money and some flowers.

After that, she proudly announced she had a new job as an air hostess and threw a massive leaving party. The new job was a fantasy, so she was unemployed for six months. She decided to go to university and started in environmental science but could not do the chemistry involved, due to her newly acquired problem with mathematics, so changed to archaeology where she obtained a good second-class bachelor of science degree. After that, she became a supervisor in a supermarket and along the way had a child. When her daughter was six months old she saw an ad for a communications support officer for stroke services at the hospital and has been doing that job ever since, a decade or so later.

Initially it was tough. Her peers had a background in speech and language therapy, nursing, or social service, while her qualifications were that she had had a stroke and recovered. She joined a community team trying to improve communication skills in those still having difficulties several years after their stroke. She was told to go ahead and make the job her own. Fourteen years later, she is still making it her own, embedded in a community speech therapy team trying to help those with long-term communication problems who otherwise might have slipped through the social service net (which has increasingly large holes).

Having been completely aphasic, twenty years later she is in a unique position to reflect on her own experience without and with speech. When she was aphasic, she never lost her capacity for thought.[4] Perhaps surprisingly, she found this a period of relative peace despite having a dense left-sided paralysis and being unable to swallow, talk, write, or walk. While she tried to make sense of everything, tapping out a few words on her laptop was enough even with poor spelling:

> Having no language was relatively secure. Starting to speak was when the frustration came. I still feel that now, even though no one knows. I didn't mind when I was German [the foreign accent], but when my vocal tone etc. normalized, I was still left with a difference between my thinking and my speaking voice. Before the stroke these were similar; afterward, my thinking voice was the same but my speech is lower tone, slower, boring, with less vibrancy or color.

Even now, though she is competent in speech, these are still present. One problem is in being succinct. Faye speaks faster than she thinks and then tries to backtrack and fails. What she wanted to be two sentences might become a long monologue in which she may not even say what she wants to say:

> I fill the airspace, it's a confidence thing. You have to be very confident to be silent. When asked a question the spotlight is on you. Often then people will fill the space with fillers in speech or just start talking and won't be what they want to say, but the spotlight is on them and their mouths are moving. I waffle and if every utterance was written down it would be very long, if just the words were written it would be far shorter.

I suggested she is on autopilot. "Yes, perfect. Social autopilot." Like most people, she presumes, she does not always think through everything she says. She finds the problem is not always in speaking, but matching words to thoughts:

> I think it's like going into a shop for a pint of milk, a loaf of bread, and butter. You go in and get a magazine, flowers, pint of milk, and forget the other two things. Why couldn't I remember those two things? Or going into the shop and not knowing where the milk is, even though you have a vision of it. But once in the shop you forget what the milk looks like but you might remember how much it costs and it's in the fridge and you go outside and remember—and then the shops shut.

Her thoughts and words slalom around like skiers, slipping and sliding across the slopes:

> I often speak faster than I am thinking. Bizarre. And then I try to backtrack and go back and I cannot. And what I wanted to be two sentences becomes a fifteen-minute waffle. My words and thoughts don't always match. My thoughts race along like a river. When I open my mouth, I am bracing for impact. . . . My words are like an uncontrollable dog.

It is in part a bandwidth problem. Listening to another person, thinking of an appropriate response, and then articulating it in real time without an appearance of effort or over-thought is arduous. She works hard to conceal that. She is also frustrated sometimes by saying the wrong thing or sometimes not having the words at all:

> I probably now express more things I am not thinking. I say something but even as I am speaking, I think I am not thinking that. I do it regularly. Whether it's getting the word wrong, or the whole thing.

At a recent conference a man was discussing the importance of psychological and peer support and how it was really important after stroke. Faye was really irritated because that is her passion too. She criticized him even though she did not mean to:

> "Why did I say that?" How many times do you think things and are glad you don't say them; sometimes my thoughts and voice go way too fast for any internal "autocorrect" to work. There is always the fear of falling foul—hence my feeling restricted career-wise.

She still goes off on tangents. While listening, thinking, and replying are effortful, dialing up the right response can be difficult too.

After one of our chats she sent me a diagram. Our chat had made her question herself again:

> I have been giving a lot of thought to our discussions and wondering why, although communication now is wonderful in comparison to 1998, I feel so frustrated so often. I have done a hand-drawn diagram that might help explain those difficulties over time.
>
> I also think that I have a constant critique in my mind now that wouldn't have been there prior to stroke. Those constant thought processes; some of my errors, maybe a lot are normal human communication, pauses, getting words muddled occasionally; something the average person would not think twice about.

The figure was Faye's three-dimensional diagram into the mind of someone with aphasia (see figure 13.1).

> My diagram, "Thought to talk," has two axes and three levels: what I think, what I want to say, and what I actually say. Horizontally, there is 1998, 2000, and 2022. Initially, thinking was a middle-sized circle, what I wanted to say was far smaller, and what I actually said was minimal. The square represents my frustration. Initially, my thoughts were unaffected; I wanted to say little and could say even less; hence the frustration.
>
> Two years later my thoughts were unaffected, but I wanted to say more. But still I was able to say very little. However, my frustration was minimal; I had a "more relaxed balance of thought to talk." Twenty years later I have lots more thoughts in my head, "more ideas over time." What I want to say is similar to 2002, but now for the first time, I say a lot and more than previously. But with this there is an increased level of frustration, a combination of things I didn't want to say, wrong words and going off on a tangent, and things I wasn't thinking.

When she reflects, though, it is not all bad:

> I do feel blessed because I had a time of self-reflection and a slowing down of life, which many may not have till they are older and which I had at twenty-one. But then I had no choice and I would not have chosen it, but I did have a time when I could stop and feel free. If I had not had the stroke, I have no idea what I would have done. I might have been a banker all my life.

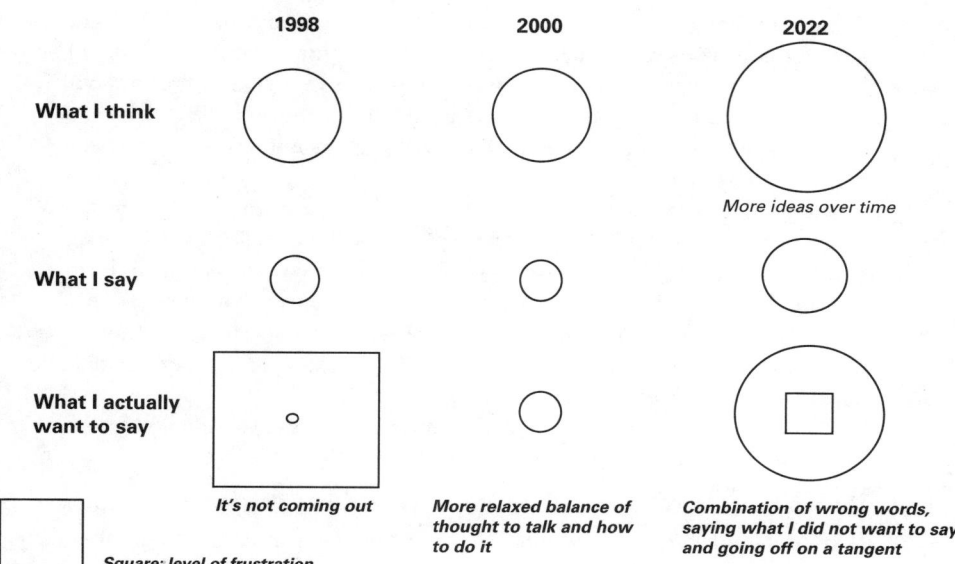

Figure 13.1
From an original line drawing by Faye Wright and reproduced with her permission.

But she also kept returning to what she called her "micro-failures," times when she said the wrong thing, when the "dog" ran off. At other times, the way she projected herself was not how she felt, in gesture as well as in speech:

> When tired I have hypersensitivity to sound. Trying to remember what you are going to say and processing what someone is saying to you is hard. This is me right now twenty years later, and still it is the case. Others look and say, "You've recovered," and I think, What do they know? How do they judge? We are having to think about how we are seen by others much more than others and all the time.

She reminded me that maybe 60 percent of people with communication problems after a stroke have depression, and though the speech and language therapists and psychologists do their best, there is insufficient support for this and other longer-term problems. Many marriages fail for this reason. She thinks that people do not realize how much even low-level support might be needed and for how long. A colleague recognizes, for instance, when Faye is tired, because her body language and speech falter. Others have no idea of the battle she lives with daily, between thinking and saying:

In the last couple of weeks I met a new social group, and I am vibrant, confident, talk hind legs off a donkey, but when we are talking and my memory is jogged and I want to tell a story and it is totally not what is in my head. It comes out and they get the gist, but I go off on a tangent, but my thoughts don't. I tell the story but not how I wanted. People see me as being fun and ebullient, but that's not what I am always.

Every day I am a different person. All processes are thought about, none are automatic. Even walking for me—I have a physiotherapist inside my head saying heel toe, lift your leg, and don't drag your butt. You need balls of steel to carry on when you have missed your point; why did you forget that, why did that funny conversation did not go to plan? But never mind, it was still okay. Another failure, for a number of different reasons, forget what I was going to say, couldn't find word. So many factors in communication.

Where is the line drawn when the words coming out of my mouth are not what I am thinking? I *was* aphasic. Am I still?

She would not have chosen this path, but it has made her question more, work harder, and—arguably—relish life more deeply. Almost unimaginably, she has been without language and then returned to tell what it was like, what it is like still. Despite her continuing problems, she has salvaged her speech and her life and, like Bill, moved from a bank to healthcare. Others are less fortunate and recover less.

14 Absence and Presence: Nonfluent Aphasia

We now consider two men with nonfluent aphasia after stroke, neither of whom has had much recovery. This discussion has been assembled from their accounts and from those of their partners.

Peter was in Nigeria visiting a daughter in Lagos when just after lunch, his wife noticed him staring.[1] They called an ambulance and then started criss-crossing the city between hospitals, from emergency departments to a CT scanner, before he was given clot-busting drugs for a large stroke. Once back in the United Kingdom, he was diagnosed as having a left hemispheric stroke, with weakness down the right side, a moderate expressive aphasia, and dyspraxia of speech, with few words and difficulty articulating them. All this was two years ago. He began to speak after five months, though with limited success.

When I visited, Peter was sitting in his chair, smart and attentive, with his wife Jean close by. He nudged her into offering me a coffee. He was drinking one with his left hand. I asked about this.

> Yes, I ccc, I can't use it un my hand. Do it all fffi fist. This is erm, um, very, I can't, I do not know the the . . . *[Jean explains that he could not feel anything with his right hand, so anything in it would drop]*. I just cannot, forget, there, gone, singnuff, fart-end. *[Jean interjects again, "Balance was also a huge problem initially, though it has improved so that he can now he could shower on his own."]* "I can't sss I can't . . . *["find the word"]*.
>
> Yes, I couldn't, try any, talk to me. I can't hear, Myst, you can hear me, I think. Yes, I can hear it. So, its in got the, quite, umm, I'm haven't got . . . *["the word"].* So, if I'm it's very, I can't hear it. There's no, if I you hear me, I cannot hear me.

Jean explained that at one point he couldn't hear—or understand?—the sound he was saying. Since he didn't know he was saying anything, he stopped trying. "I kept telling him to shout so we could hear him." He can copy and write his signature but otherwise cannot manage more than a few letters. Reading, similarly, is possible for a few words only. He taught himself to use an iPad and will start emails and then give them to Jean to finish. Now he also consults the financial pages and accesses their bank accounts. Jean keeps tabs on things.

Yesterday he had been to a meeting of a retirement organization. "They're good, they wii, w, would . . ." ["Talk to him"]. "Yes, they do. I can do it." But he cannot follow a whole lecture. "Yes, I cannot follow, next, I cannot, I haven't got . . . I for, I for." Jean interrupts that he forgets. He likes the TV news. She says he had such an active mind; she wonders whether he's bored. It's clear that he is concerned about her too. "Um, I, I can. But it's hard for you, because I can't. Don I. Because I forget."

He goes off for walks tagged with a tracking app. At the village shop he can buy things if Jean gives him a list. He also uses his charge card. He remembers the positions of his PIN numbers but is pretty safe with it because he does not remember the numbers themselves and cannot say them aloud. Peter laughs. He gets up to move to a table and, from a drawer, hands me small, laminated lists of his prompts.

Places to eat

The Crown, Carpenter's Arm, New Queen, The Jetty

Local places to go:

Shops

Post letter parcel delivery next day

Numbers

Up down in out

Where when

Yesterday and tomorrow, week month year

Relatives, family names.

He shows me his prescription drugs: anticoagulants, statins, antihypertensives, and a stomach protector. "Welcome to your prime," I suggest, and we both laugh. Jean explains that he had a heart attack after the stroke, following a previous one at age thirty-five, and also had a stent inserted twenty years ago. I ask if he'd like to say far more. "Yeh." Frustrating? "Yes." Were you angry? "Yes."

Jean adds that they both had to get used to their frustration. "Finding a position in bed was difficult with his balance. Some nights I stayed down here and I thought, 'At last, on my own,' and I think you liked it too being on your own. I think we've both got over that. Though occasionally the frustration of not understanding is still there."

I wondered what Peter liked to do.

> but um. I can do, I can do the walk, long. I can do that. She hasn't got the . . . [Jean also has heart problems so cannot do hills.]

Jean explained that if one of their (adult) children phone she'll hold the phone to his ear while they prattle on. Peter adds, "It's good." Jean carries on: "Good for them too. My daughter was quite affected. The grandchildren just come up and chat. It's just Grandpa. They accept less language."

Our chat has taken a disproportionate amount of time and I thought Peter might be tired, so I thanked them both and left. Later, I arranged to talk with Jean on her own. I began by suggesting that living with aphasia is about the partner and loved ones as much as the person. She responded:

> Absolutely. If I ask him about when it happened, it's a blank for him. It's not just a reduction in language; it's more than that. Life is turned round completely. He was very much the talker, lots of energy, always doing something, loved his work. I now do all the talking with friends. I am always putting on a good face since no one wants to see anything else. All the decisions, say, what paint color for the bedrooms, are down to me. He will say, "What's your feeling," and then say, "Yes."

He self-cares completely, though initially Jean had to help. That was fine with her; after all, she was a nurse:

> He was ill, it was a shock, but I just looked after him. Then it hits you that this is going on forever. He will very occasionally initiate a conversation but otherwise nothing. We may have plans for the day—I may have plans—and then he says, "Are we going out?" or that is what I think he says. He thinks I have all day to look after him and entertain him, unaware of the work I do to keep it all together.

He finds social situations difficult and will hide away upstairs even from the family: too much noise, too much information.

> With the family, he may start and you can understand the first few words and then it drifts off. I have to try to work out what he is saying. That is the worst bit of it all and you think you are going mad. You've not only got your thoughts, but his too which I don't understand.

Early on they would get so frustrated with each other that they would have to stop:

> We went through a phase which was really awful and he would get upset and angry and would lash out, once hitting his head because it was not working properly. When the atmosphere was dreadful we would part, him to his study upstairs and me downstairs and then after a while we'd carry on. Now it is easier, we know our limits and have accepted them.

All unsaid, of course.

> Sometimes I feel I'm a carer, other times a version of what was before, not quite the same as before but something.

Jean can now leave him for a few hours. They use an alert and both know if something happens, he can press that; if they didn't have the alert, Jean said, "I could not even buy the bread."

The aphasia has completely inverted their relationship. She now has to make all the decisions and finds it—day in, day out—so boring. He cannot make choices. "I try to get him to reply yes or no, tea or coffee. Sometimes it drives me insane while at other times I can ignore it." He doesn't express wants. He makes tea but never says he wants one or what he might like for dinner. "I don't think he cares what he has so long as it's in front of him. Early on I would go to shop and came back after lunchtime and he was still in the same place watching TV. Had he missed lunch? I've no idea. Can he care? Does he have the desire?"

> I chat to him but we cannot chat. When I chat he looks at me with a vague look. Sometimes I hear myself chatting and think, "Why?" I chat now with friends. I have to think about what I say to him to make it simpler. If without thinking, I say we need to something, such and such, it may not go well. We've lost the key to the garden shed and I needed to put the bulbs in. Next time I was out there was lots of banging and he'd taken the door off the shed so I could get in. He's helping but not the usual way.

Peter retains his sense of direction and goes on the bus into the local town and then takes another to the city. He always comes back. I picture a man walking in a big city, alone, silent, unable to talk or to ask anything of anyone. Why does he want to go? What is he thinking?

Jean wishes someone had said how much care and how much independence he needs. He wants her to visit the family and he'll stay at home. He'd probably be okay, unless something happened, and then he wouldn't.

I still understand some of his presence, but the blankness is quite scary and then I feel I am just looking after him. He cannot express what he's thinking or feeling. I wonder if his feelings are going round in his head and trapped but I have no idea. I just feel sad for both of us that this is where we have ended up. Back two or three years we had such a different lifestyle and now it has gone. When I do see someone it is quite a shock to talk, though I've got friends and children with whom I natter. People you see every so often we've had here, and I feed them, and when they've gone I wonder if we'll ever see them again since they've only listened to me chattering. Previously, Peter would chat away and it was completely different. He was a presence.

Every time he speaks he has to make an effort, and then the timing is off. Sometimes silence and you expect him to say something and he doesn't, so I cover.

I feel guilty if I go out. Shopping I feel guilty if I look at clothes in another shop. The other day my hair appointment was delayed so I had a coffee and I felt guilty, even though he was perfectly okay when I got back. You want someone to say it's okay to do that.

I replied lamely that none of this was her fault. But, of course, she knew that. A Church of England vicar once said to me that during a period of huge emotional turmoil for him, with his stiff upper lip he had no one in whom to confide. So, he said, with English decorum, he might "drop the tea tray." Jean, similarly reserved, might want to drop the tray, but her concern might be that no one would hear it.

England 3, Senegal 0

Ron and Rita's life revolved around dancing. They would go three times a week every week. He had been a printer in London before they retired to a bungalow in a cul-de-sac on the south coast of England. There, when they were not dancing, they'd be walking either along the local beach or further afield along the coast. For holidays they would take cruises, at least two per year.

Last summer Ron had a severe stroke. Despite having prompt treatment with removal of a carotid artery clot and stent insertion, he could neither talk, write, or swallow. He had nasal feeding for two weeks, but after that he was back with Rita. They had also found lung cancer and drained fluid from his lung but could do no more.

Once home he still could not swallow easily, so she'd liquidize all their food—he'd only eat it if she had the same. Fortunately, they had already downsized from the bungalow to a retirement flat in a small complex.

When they first moved in, they were always out walking or dancing or just exploring. Now Rita did not leave him for a month. Then she managed to nip to a supermarket and had her hair cut. When she got back, he'd made himself a sandwich:

> He dresses himself and personal care and eats what he likes, small portions. Walking he needs a stick and a frame he can use as a chair. We've sold the car—he'll never drive again.

I ask Ron how his speech is and whether he can make himself understood.

> Some, sometimes, yes, but if I know say what I am saying if I want to see anything and sometimes, I say yes and then nothing at all. See what, I don't really, I don't say what it is, can't certain things, yes.

Words remain but meaning is just out of sight, for me at least. Rita does not think he understands what he is saying. "He never asks a question and rarely answers one."

Simple questions seem easier. He agrees, though his answers often tail off:

> Yes, what was it you ask for? Sometimes I got to ask again to see what it is, yeh, for me anything evenly, I don't know what I'm talking sometimes, talking I can't alter it, say sorry I can't do it, just say it all over, it's gone.

The football World Cup was on, so I asked if he liked football. He could remember who England was playing next and their score from the last game: "Senegal 3–0. I enjoyed that. We had it most television I would see the actual hopefully, yes." The medical care team had suggested Ron go once a week to the hospice to give Rita a break. The speech therapists also offered her hospital-based sessions. But none comes with transport, and a taxi would be too expensive.

I asked Ron the best part of the speech therapy:

> Well, the thing mainly or er or like now I start I don't really ask in my times something there are better than what you are doing. I don't know you can't really give I would not have a job to phone if sometime if up the road and back in and leave the message and I wouldn't ask, I could not really say who it was.

At times he is close to expressing himself. I asked what his job was.

> A printer. I used to be, my friend said things like that. We went into an F about it, instead of ordinary slip, it's gone into a passive, I can't think what I was doing.

I asked if it was frustrating. "What?" "Now?" "Yes." He is angry with Rita sometimes, especially in the evening when he is tired and his speech deteriorates. He enjoys watching his favorite old programs on television, though Rita won't have it on during the day.

Despite downsizing, Rita explained, their rates (of local amenity taxes that had bands according to the property's size) had gone up from an E to an F band. For what? The bins emptied once a week early in the morning, right under their flat with all the noise and fumes. "It takes ten minutes maximum. That's all we get." Ron joined in too:

> No. Went back in the in the F band, before we much bigger what we had and we only had an E rating, a much bigger place. We did not know then [when they moved in], did we?

She worries that he might be in pain from the cancer and not be able to tell her, and she's upped his medication once. I mentioned that I knew at least two people with aphasia whose partner had become their carer. She replied with acceptance and with a quiet and moving dignity: "I took him in sickness and in health, and I will look after him, that's my job."

It was, however, something she had said to me when first I arrived that lingered:

> In the morning we'll be lying there together in bed, he'll talk, and I don't know what he is saying. Life cut short and dead. Ron eighty-five and I am eighty-nine but we were still active. This has been very hard. No more dancing; he could not remember the steps.
>
> You would not believe he was intelligent before. It's like half his brain has disappeared. You don't realize how difficult it is till it happens. No one can realize till it happens.

How we express ourselves, and particularly the words used, is a large part of how we are perceived, and without intelligible speech people might think intelligence itself had been affected. Yet one should be very careful before reaching such conclusions. Sheila Hale quotes the distinguished neuroscientist Antonio Damasio, who has collaborated with people with aphasia for many years:

> As I studied case after case of patients with severe language disorders caused by neurological diseases, I realize that no matter how much impairment of language there was, the patient's thought processes remained intact in their essentials and, more importantly, the patient's consciousness of his or her situation seemed no different to mine.[2]

More recent, large clinical studies have shown various results. In a large meta-analysis of 47 papers and 1,710 people with aphasia, Fonseca and colleagues found that on tests of nonverbal reasoning, working memory, and nonverbal memory, those with aphasia performed about the same as those with similar strokes elsewhere in the brain, suggesting that aphasia did not affect these cognitive functions over and above the effect of stroke in general.[3]

In contrast, Hanane et al. found that in 147 patients with aphasia of mixed types, assessed up to a year post-stroke on reasoning, visual memory, and executive functioning, there were deficits. Those with continuing aphasia had lower cognitive domain scores, worse functional outcome, and were also more often depressed than patients who had recovered.[4] In a study of 189 patients with severe global aphasia, Marinello et al. found spared cognition but deficits in visual and auditory recognition.[5] More recently, Crinion, Leff, and colleagues analyzed language and nonlanguage cognitive tests in thirty-six adults with long-term speech production deficits from post-stroke aphasia. All participants had difficulties naming things, with relatively intact speech comprehension and no apraxia of speech. They found impairments in non-language-specific cognitive function. They suggest that language processing should be seen within a general context with perception, action, and conceptual knowledge being linked and mutually interactive.[6]

The argument rests on whether language is a separate domain in the brain, vulnerable to damage that leaves other functions intact, or whether language is connected with and parasitic on other domains and other brain areas. Those with aphasia appear to retain a good knowledge of the world, so their semantic systems remain, and they do not have so-called thought disorders like paranoia. But if you look in more detail and depth, other subtle deficits can be observed, say, in thinking and frontal executive tasks, for example, decision-making. Some who think there are separate domains within the brain would say any subtle deficit beyond language with aphasia might reflect additional system damage, while those who think everything is more linked and interdependent would simply point to results showing just how rare are isolated speech and language deficits.

Be that as it may, with Ron the main thing seemed to be to presume he was interested and intact and to engage with him through speech and any other means. His interior life was unclear, but whatever it was, we needed to try to reach him as best we could. I hoped Rita might see how intact Ron's thinking was under the aphasia, and that she might, perhaps, get more into football.[7]

Oliver Sacks once wrote of some conditions being neurological hells. But some conditions appear not as hells but as more liminal, marginalized states with presence reduced, or concealed, or minimized, with little communication possible. As we have learned from Fay, from Bill, and from Steve, many people with aphasia are able to return from their wordless state to reclaim speech as best they can and, with that, their identity. Though sometimes altered and needing support, they are back. But many remain stranded by an aphasia and disconnected from their previous social existence. While they may—one hopes—see and hear their partner's love and support, it can preclude their showing their love in return, further diminishing their presence. Less a hell than a living limbo, barred not from heaven by being unbaptized, but from a normal human, social existence by the effect of stroke on speech, with their past presence replaced by an almost unfathomable social absence in which their interior lives remain unclear. And their loved ones, often older, isolated, and alone, also live with their own consequences of aphasia.

15 Eloquent Rhythms of Nothing: Fluent Aphasia

Whereas Broca's aphasia, as seen in chapter 12, particularly affects speech production, in fluent aphasias, speech persists. These are subdivided, as we have learned, into those in which language is largely understood, conduction aphasia, and those in which speech and language understanding is impaired, whether the speech of others or of themselves. This chapter considers two people with different types of fluent aphasia, one with Wernicke's aphasia and the other with a more unusual response to his loss of speech after a stroke.

Whole New Normal

Linda walks the dog first thing while Paul is still in bed.[1] "I have to get out. I may end up an old dear walking round the park talking to themselves or to the dog. I talk a lot to her [the dog].[2] If the carer comes I go out and be back to make him lunch and find him a TV program." During lockdown Linda and Paul would walk in the park each day, but now he can't be bothered. His walking is not so good anyway, so he just watches TV while Linda finds something to do, though often he'll call her for something or other. In the afternoon she feels bad leaving him on his own for too long, so she'll make tea, sit with him, and chat, while the dog snuggles up to him on the sofa and he makes a fuss over her.

Paul is usually up at ten o'clock, when the carer comes in, and has a shower. He can dress himself, though he can be confused about what bit of his arm or leg goes where—a dressing apraxia. He can do shirt buttons, though his right hand is clumsy. Sometimes he'll shave with his toothbrush. Once dressed he'll go downstairs and the carer gives him breakfast.

After that she may stay for two hours and help him with his hand exercises, since his right hand is bad. Left to himself, he'll just watch TV. He's fine, though if the routine is altered, he can become confused and then panicky.

Sometimes in the afternoon, they go to a movement class run by the British Gymnastics Association for those with a stroke, dementia, or Parkinson's. It is a seated class for both fitness and stretching, but they also encourage hand and finger exercises. Afterward, they sit together and chat. They talk too fast for Paul, and if someone asks a question Paul may start to reply, but then not make himself understood and stop again.

If there is no class, then he'll just watch TV, with Linda spending the day channel-hopping to find sport or some other programs he might like. Later Linda prepares a meal and will help him feed if necessary. He'll go to bed around 10:30 after the news, though it can be earlier if there is something on TV he cannot follow. If he enjoys a program, he is engaged; otherwise he falls asleep. He loves sport, especially rugby and cricket, and appears to follow it, even though his sight is not great.

Four years ago, he collapsed in the bathroom up against the door, so when they arrived the paramedics had to call firefighters to bash it down. He'd had such a large stroke that at one time doctors thought he might not pull through. He spent six weeks in intensive care and then the stroke unit, before nine months in a nursing home hoisted to and fro, with people assuming he'd need full care from then on. Then one day they found him in a different part of the room—unbeknownst to carers he'd started to move. Physiotherapists were brought in to help him with mobility and he relearned to walk, while speech and language therapists reappeared too (they had tried before and given up).

As he improved, his voice returned. Linda remembers, "He'd do one word and I thought, Brilliant! but then it became apparent that as he tried to explain what he was thinking, he could not make himself understood." Paul produced longer and longer streams of unintelligible words and sentences:

> All those I ber simple seredimple, sererosamemoru. It was comidram. I could say slas, that builded, Ican very different feel, if I gave a word wissage of all ris aris, too russle six I preclusid all thse people parcity, I kept my personally.

Paul had a form of Wernicke's aphasia, and though he found it difficult but not impossible to understand the meaning of spoken words and sentences (and reading and writing), as speech returned, though he produced sentences with phrasing and expression, the words often did not make sense, something of which he seemed completely unaware.

I had first met Paul at a group run by two speech and language therapists in a small, forgotten part of a hospital for those not coping with aphasia further down the line. They run a ten-week course of two hours per week, trying to reach out to a small cohort of patients, who are often older and years after their stroke, and often alone and isolated. The group had three clients. One was George, in his late eighties, whose sole friend had been "gradually disappearing" and had just gone to a home for those with Alzheimer's. They were keeping an eye on George and hoping they might find him a suitable social group to join. In contrast, Jacqui, in her seventies and three years since her stroke, appeared quite fluent at first but it soon became apparent how flimsy her speech really was. She found coping on her own bewildering.

Paul, in contrast, retained his desire to utter and was fluent in speech, which in turn showed a rising and falling prosody, accompanied by facial expressions and gesture; it was just that his words were meaningless, of which fact he appeared unaware. His command of prosody and expression was so convincing that Jacqui and George thought they might be understanding him, but this was based on his fluency rather than any meaning conveyed in his words:

> Can't make cobis of it I could use words coastload hopicing there you Europe copitate vertical them. But you can't locals it, lumbering one taper to members consciously tries to blow on one to one to eight massively easier.

After several years neither therapist thought improvement likely and had invited Paul for a second course simply to allow his wife Linda some time off. I approached Linda and Paul to ask if I might visit them at home and did so several times, and also chatted to Linda in a café away from her caring duties. Paul had been an officer in the army and a high-up Mason. Until his stroke he had been used to words, to oratory, and to commanding—indeed, at times dominating—a room.

Now Paul could use everyday phrases such as "Please" and "Thank you," appropriately, and when the carers left could say, "Bye" and wave. Otherwise, he could communicate little through speech. He was interested in my coming to chat, and intent on listening and followed much of it, I felt.

Linda explained that he liked having friends round but may not follow conversation if it became too fast, with too many strands; then he disengaged. I asked Paul how much he could make himself understood.

> Good question. I can Farook, sinjst, you are gootst me you know and I canotdisssy dissy this. Sijjer, its cor, one klike judder, no it doesm ehte one wafer, might be lovely at moment, I couldn't . . .

Though initially present, sense was soon overtaken; sliding into fluency without meaning was far easier. Linda thought there was so much that Paul could not express: "I can work out what he is saying a lot of the time. At other times I can be completely wrong and he tells me." I learned that Paul's answer of "Good question" was a frequent reply before he reverted to incomprehensibility. Linda suggested he needed simple questions. "If you ask if he wants tea or coffee, then he cannot decide or answer. You have to ask if he wants tea and wait. If not, then does he want coffee? Then he is fairly reliable."

Linda asked Paul whether, when he addressed her, he knew when it had gone wrong. "What do you mean. . . . What I'm saying but it's not laying. I say no, yes." Linda continued, "So you know we don't understand." "That's right." She asked him how frustrating this was, thoughts without expression. He laughed, "That's right." He could not write either; his stroke had affected hand function and writing. His eyesight had also been affected over twenty years ago when he had had a heart attack. He had been over-anticoagulated, leading to a brain bleed that left him with a hemianopia. He still lost the end of words and needs large print to read anything. "I can see them whister, as if it's silly, I need get perforsan do it again, do it again the fast finnin lid, I mamamori." Then he stopped as though aware of his speech and said, "Let's start again." I asked if he could understand me. He was sitting so attentively, listening and responding, both to direct questions and sometimes on his own. "Yes. Good question." I asked, 'Do you answer me in your own mind? "Yes." I continued, "Are you aware I cannot understand?" "No."

He is a Mason, loving the ritual and structure, which is a little like the army. He had also enjoyed being in charge too. After the stroke he managed to go back, which allowed Linda some evenings to herself. He would say it was "Okay, but I cannot do anything." First, they cut it down to one lodge once a month, and then he stopped. He cannot take part and found the dinners difficult without someone to help feed him. They had to order non-alcoholic beer for him too, since once he drank too much and had continence problems. But the main problem is being unable to make himself understood. "This is one difficulty. I cann, memmorit, went to the bitten, I went to the not the, God, I couldn't do its . . ."

Linda explained how much he hated if the Masons got any of their rituals wrong. Again, Paul started to talk, words fluent but meaning hidden. "We did useless only, get out adamant. Cannit remember words, cannot say them. Sometimes I can't word it. I've been along, It, it it, can nose that sort of sojit. But it couldn't say."

Later, Linda and I chatted in a café during one of her precious mornings off. It's clear, she said, that he has recovered some slight understanding and speech but that this is not very useful:

> He was so articulate before. That he cannot communicate must be terrible though his talking has improved a little. Initially, he would give stream of words, unaware he was not being understood. He used to get very angry, less so now. His daughters used to think he was always telling them off. He kept ranting on as though asking why they could not understand him. He'd become very frustrated and that would upset them. I got it a lot.
>
> Now, four years later, he is calmer. He'll try to explain and the first part of the sentence may be fine but then he gets into difficulty and it trails off and we cannot understand. Then I'll say I cannot understand and he'll try again and then after a couple of times he'll say, "It's no good, it's not good. No, no, no, it's not that," and say something else. He realizes people cannot understand and with more effort he can start a sentence, but it gets to a point where it either runs away or more often he will stop and say, "Forget it. I'm stopping."

She worries that he may have things he'd like to tell her but cannot. For instance, when his ear or teeth hurt, he points—but not always to the right place. She has to ask if he means teeth, or does he mean ears?

He has actually lost more than speech, suggesting widespread damage to the frontal executive areas of his brain too. He might leave the lounge to do a pee and then forget. If he has an accident, he is not upset. He'll not go to change his trousers and may become aggressive if it's pointed out. He cannot work the TV remote, so if he wants something changed, he makes a noise and Linda has to know what he wants. He's also become more self-centered, says Linda:

> Trying to say what he wants, he whines. He also whinges a lot. I find it very annoying. He was not a whinger before; he was a sulker. He lives in his own

world of "Me. Me. Me." My needs are not on the map. He is always demanding something. If I explain I am busy and cannot do it now, then he does understand.

Now he can slightly self-monitor and know he cannot be heard where previously he was not aware of his errors. At first, he presumed what was coming out was correct.

Before this they were traveling quite a bit and had more plans. Now, "It's a whole new normal." They have lost friends and acquaintances, who visited at the beginning and then fell away. It can also be a problem for the family. Paul has three daughters, all of whom live a distance away and have their own families. Initially, they found it very difficult to cope with the change in their father; now one of them comes down regularly once a month for a weekend to allow Linda a break and maybe visit her son for the weekend. The other two find it a bit more difficult but will come and sit with Paul during the day.

Living with Paul requires cunning. He likes alcohol a little too much, so Linda dilutes all his wine, rations him to one glass per night, and keeps the drink out of reach. Once in the early days he was drunk and she had to call an ambulance. They were unsure if he had hit his head or not, so he had to go to the hospital. When Linda went in, "The doctor said there is good news and bad news. He's okay, but he's ready to come home." He will eat anything he's given and does not express desires (beyond alcohol). Shopping has been a problem. He sometimes wants to go to a supermarket and buy things he shouldn't, like pork pies, snacks, and chocolate.

She thinks it might have been better if he was in a wheelchair and able to make himself understood. "Before the stroke, I used to moan because he would never shut up. Now I'd love him to talk. We cannot chat. He watches TV endlessly." In the army and the Masons, Paul would talk and talk.

If Paul could use six words instead of one he would; often words they understood but I could not. Now, I still cannot understand what he says, but I cannot look his words up either.

His situation and those of others with this form of stroke raise questions about the interactions between words and their embellishment and embodiment through prosody—the pitch, loudness, and intonations to stress certain parts—and the gestures that surround them. While language is normally a complex symphonic dynamic unfolding, Paul's aphasia largely preserved much of this dynamic but robbed him of the semantic

component. Since his words are bereft of meaning, the prosodic and gestural components lack context and meaning themselves. Yet so used are
we to listening to speech's performative element and its gestural accom-
paniment that we would strain to discern meaning from his utterances,
beguiled by his tone and prosodic skills and thinking it was our fault we
could find no sense in his words. We now turn to a master of this expressive
art without words.

Still John

One of the most extraordinary accounts of a fluent aphasia is not from
an eminent neuroscientist or clinician but from Sheila Hale, a writer, art
historian, and journalist whose husband developed it after a stroke. She
explored the condition, turning herself from a novice to an expert, as she
wrote an account of his life, *The Man Who Lost His Language: A Case of
Aphasia*.[3] A writer herself of eloquence and renown,[4] she is well-connected
and fearless enough to approach many of the main scientists, doctors, and
speech and language therapists in the field. An additional strength of her
account is that it unfolds over years and focuses not only on her husband's
initial stroke but on his recovery afterward, his rehabilitation, and how he
lived and came to terms with his new situation as best he could. She was
a researcher but also his wife, carer, biographer, champion, and advocate.
Such extended accounts of neurological conditions are as comparatively
rare as they are essential.[5]

Sir John Hale was no average individual. Deferring his place at Oxford
until after the war, he served in the Merchant Navy, since osteomyelitis
prevented him serving in the Royal Navy. Once at university he took to
acting and was sufficiently dashing to be offered a part in a movie opposite
Jean Simmons.[6] He turned it—and her—down for academia, staying on at
Oxford as fellow in modern history at Jesus College until 1964, when he
moved to become founding professor of history at the then new University of Warwick. In 1970 he was appointed professor of Italian at University College London,[7] where he remained until his retirement in 1988. In
the United States he was visiting professor at Cornell and the University
of California and research fellow at the Institute of Advanced Studies in
Princeton. He was also a fellow of the British Academy, member of the
Accademia dei Lincei, trustee of the National Gallery London 1973–1980

(chair from 1974), trustee of the British Museum, chair of the Government Arts Collection, and member of the Royal Mint Advisory Commission. He was knighted in 1984 for services to scholarship and the arts and awarded the Italian order of *cavaliere* (knight). Along the way were numerous books and articles, and he also appeared frequently on television and radio. His only hobby listed in *Who's Who* is "Venice."

Then came the stroke, with right-sided paralysis, temporary swallowing difficulties, and an enduring fluent aphasia. Once the acute phase was over, Sheila documented his abilities, his losses, and her own feelings and thoughts. It was

> an essential article of faith that John could think as clearly as ever and understand what people said to him . . . I could accept the loss of speech because the fire in his eyes, the lively modulation of his voice and the way he shaped it with his left hand told me that John was still John. I could accept he didn't seem to recognize the words for razors and pencils and keys and the like, because he was so quick to respond to complex ideas.[8]

If his thoughts were intact, then so was the penumbra of vocal intonations and gestures woven around speech. An educator and lecturer of rare eloquence as well as a skillful committee man, his life revolved around erudition and conversation. Much of this remained:

> A psychologist specializing in pragmatics might take note of John's normal—indeed gracious—turn-taking in conversation; in the way he can use silence to convey scepticism or surprise; in the way he points his finger to indicate *here* or *there*; in the body language that looks so normal from a distance. . . . He can calculate very precisely the effect of a gesture, the length of a gaze, a pause, a lift of an eyebrow, the position of hand or fingers. He can modulate vocal tones to convey the full range of crucial sub-texts - extreme courtesy, irritation, humility, sarcasm irony, menace and so on—without which spoken language loses much of it meaning.[9]

For John, talking was not hard. Sheila wrote of his continuing relish for conversation and of others congregating toward him as moths to a flame:

> We are still invited to dinner parties, big parties and, recently, to a ball. No sooner have we arrived than he leaves me in search of interesting conversations, . . . greetings friends and strangers alike. . . . I find him surrounded by laughing people. They insist he is as good a conversationalist as he ever was; but they want to know more precisely what he is talking about . . . People who loved John before his stroke love him more now. There is something about his eyes and body language that seems to speak more directly to their hearts than all the words with which he charmed and taught throughout his speaking life. Strangers are

fascinated by this social, able stranger with the donnish manner who listens so
intelligently but replies in an unrecognisable foreign language.[10]

Later she writes of how "John's manner and voice are so obviously pregnant with meaning—and he so obviously believes he is communicating
meaningfully—that even Elizabeth (the speech and language therapist
he liked) found it hard to credit her own evidence that his voice was not
describing real words and sentences." He had "begun to attract a reputation
for vatic wisdom. After a friend's lecture he stood to ask a question . . . The
learned audience seemed no more puzzled by his discourse than they had
been by the lecture that prompted it."[11]

Speaking, albeit in persuasive fluent nonsense before an audience, was
good for John himself:

> When John talks he feels at ease, as though he were communicating perfectly
> well. His conversational behaviour is so natural that many people are persuaded
> his voice, although evidently naked, is somehow also clothed in meaning. . . .
> People tend to blame themselves for their failure to understand him. John's
> charm and thespian qualities, underscored by his expressive voice, make him not
> just socially acceptable but an object of fascination. But his confidence is also the
> most worrying of his symptoms. Specialists . . . have advised me that John will
> never regain his speech until he recognizes that he is speechless.[12]

All the more remarkable, then, that John's form of aphasia was not a
classical Wernicke's. Unlike Paul, he could use only one utterance, made up
of two or three syllables. Sheila noted:

> John only had one utterance, "Da Woah," which he would sometimes break up.
> He used two syllables to put forward the rhythm of a sentence; it had all the
> prosody of a normal voice, and he only had his left hand [to gesture, due to the
> stroke affecting the right side] to do mimes and illustrations.[13]

Usually, Broca's aphasia is the opposite of fluent, with the few words that
remain delivered with great hesitancy and effort as though almost painfully
awkward. John's stroke had apparently robbed him of his vocabulary but
not his abilities to converse with prosody, making his social performances
and speech acts even more remarkable. He combined a Broca-like paucity
of words with a fluency of speech and an understanding of others' speech
but without any awareness of his own limitation, which suggests a form of
conduction aphasia. Given the combination of extraordinary fluency with
absent vocabulary, some of his problem may have been a speech apraxia as
well as the aphasia.

Sheila not only observed his behavior and the skill with which John talked, but—a far more difficult task—tried to understand what he might have been thinking without verbal language. She itemized how he could not unravel the discrete element of a sentence, like a verb, which might be ambiguous:

> I could believe that in John's mind the world was now represented not in sequences of abstract signals but in images that stood for aggregations of the elements of sentences. For him the packaging of a pair of socks displayed in a shop represented the way I would see the socks when I went to buy them for him. If he wanted to refer to friends, he would draw an accurate map of the way to their house from ours. If he wanted to tell me that the doctor had come while I was out he would draw a man taking another man's blood pressure.[14]

She explored how much might be carried in communication by the words and how much by gesture and prosody, quoting the composer Leoš Janáček:

> When anyone speaks to me, I listen more to the tonal modulation in his voice than to what he is actually saying. From this I know at once what he is like, whether he is lying, whether he is agitated. . . . I can even feel, or rather hear, any hidden sorrow. Life is sound, tonal modulations of human speech.[15]

Remember though, John's "vatic wisdom"; without words, much of the underlying meaning and information in a speech may be lost. Its modulation by gesture, phrasing, and prosody may improve eloquence, but words convey hard information. Here a distinction might be made between speech as information transfer, as in a lecture, and speech during social interactions. In the former, words are crucial and might be read as well as spoken (indeed, some philosophers and others do read their papers out loud word for word from their script), while in the latter they are accompanied and enhanced and complemented by gesture, prosody, and softness of tone. We constantly judge how personal and full of feeling our speech and its embodied accompaniments should be and alter it to social situations, as do politicians, for example, who shape their speech and accents to the audience with an actor's skill.

Perhaps most of all Sheila was fascinated by John's thought and its relation to language and by her "desire to share the experience with John, to know at first-hand what it is like to be him." She quoted Einstein, and then Levelt:

> Having arrived at results perfectly clear and satisfactory to myself, when I try to express them in language I feel I must begin by putting myself upon quite

different another intellectual plane. . . . The words or the language as they are written or spoken, do not seem to play any role in my mechanism of thought . . . The desire to arrive finally at logically connected concepts is the emotional basis of this rather vague play . . . [which are] in my case of visual and . . . muscular type. Conventional words or other signs have to be sought for laboriously only in a secondary stage, when the mentioned associative play is sufficiently established and can be reproduced at will.[16]

There is no single language of thought . . . The cognitive system communicates internally by means of a number of conceptual codes: propositional ones, spatial ones, kinesthetic ones, and probably others.[17]

She questioned herself and her assumptions, and indeed those of the great figures in linguistics, as she was drawn again and again to the dominance of language in our experiences of others:

Most of us can't help judging the quality of thought . . . by the language in which it is expressed. . . . The folk metaphysics of expressive language as a prerequisite of thought has been embedded in philosophy at least since the ancient Greeks used the word *logos* to signify both thought and language. . . . Saussure was categorical: thought and language were inseparable, interdependent and simultaneous. . . .

Consciousness and language are intricately interwoven. But are they really one and the same? . . . Even the most profoundly amnestic patients on record have not lost language. . . . Alzheimer's patients in the last stages of dementia, when they may no longer know who or where they are, often retain the ability to speak in well-formed sentences.[18]

She believed that John's spatial and kinesthetic thought processes were intact, that he believed he thought in language and that his problems "are not in pre-verbal messages but in the processes by which it is translated into communicative language," as Einstein had articulated.[19] To satisfy herself and to allow John to reveal his continuing intellectual prowess, he undertook a series of tests. When John took the Standard Aptitude Test (SAT), which is used in the United States to test education and intelligence, he reached a level commensurate with a scholarship candidate for Harvard. He did well not only on nonlinguistic tasks, like spatial relations, but also on vocabulary, though less well on grammar. One friend, an attractive woman, once gave him some Scrabble letters, EOSR, wanting him to rearrange them into ROSE. Instead, he mischievously spelt out EROS. He knew what she wanted but couldn't resist. He developed some ability to write later and would send short postcards. He had a collection of postcards of paintings that Sheila would show him; the first word he wrote after the stroke was "Vermeer." His ability to recognize painters appeared undiminished. Sheila was satisfied that John's cognition was largely intact, though of course this immediately raised questions about how he had learned to live inside the condition.

She wondered about John's inner voice—quoting Linda Wheldon, a psycholinguist—that speech has phases. "The prelinguistic stage is called 'the message.' People are very skimpy about what this may actually contain, i.e., what kind of units. . . . The second stage is grammatical coding: you retrieve the lexical items that best matches clusters of concepts . . . and build a syntactical structure which puts them in the right order. The third stage is to generate phonemes, the sounds that make up spoken words." She wondered—hoped?—that John may be intact at first stage, but then with overload, selective attention, and problems with monitoring speech, "like proof reading," it became difficult not only to talk but also to monitor what he said, so he could not, for instance, repeat what is said. John apparently could not monitor his own speech despite understanding others and thought that the ordered sentences in his head he spoke out were normal as ever.[20]

Sheila tested this on a single occasion. She told me:

> I once taped his speech and played it back to him. The worst experience of his life. He had no idea. It was as though I had stabbed the person I loved most in the world.[21] If he was fully engaged in the subject of the conversation and the other person was intelligent—military history in Shakespeare, which interested him— then if you made it clear that you could not understand, then he would say his "Da Woahses" very slowly and clearly, so they might understand.[22]

To present him with this, she suggests, "is a ghoulish nightmare." At some level, he could recognize his speech and be horrified, even though what he heard normally, or believed he heard when speaking and apparently (mis-)monitoring, was so different from what he heard in his own inner voice.[23]

As she went closer and closer to John's inner world, without being ever quite able to enter it, she also gained an inkling that these fluent though meaningless words were not entirely worthless. He could listen to himself say, laboriously, his few meaningful words. But if during a dinner he tried this, he soon reverted to his usual fluent performance, "eloquent rhythms" of nothing (in word content at least).[24] She wondered if he was in denial of this or unaware at some level. Note that he, like Paul, could say a few words with meaning and respond to a question but that speech soon tumbled into fluency and the meaningless. Fluency allowed John and Paul to speak, to express themselves through their eloquent use of variations of loudness, intonation, and gesture, more comfortable in their fluent worlds than in ours. In one passage, Sheila observed that "It might help him more if friends were less ready to interpret his non-verbal language."[25] By tacitly encouraging him to proclaim at cocktail parties in the soirees that were his social lifeblood, people may have been encouraging him to remain in that world. She saw how much John enjoyed cocktail parties, "even though the only speech he does not understand is his own."[26]

If fluency was social and a continuing presence, and his few meaningful words stuttered out with considerable effort and anguish, then one can see the appeal of the former for John and for his friends. How many of us listen attentively to friends and relatives telling their favorite anecdotes for the umpteenth time and never point out we have heard it before? This is not simply politeness; we may enjoy the delight the speaker takes in the telling. Does listening to someone with a fluent aphasia not contain some similar compassion? Sheila was less convinced. She knew that "until and unless John listens while he talks, I will never again have a real conversation with him."[27] Moreover, she reminded me how much John himself wanted to speak intelligibly once more, that at some level he was aware of the fluency trap and wanted his communication to be normalized.

A year after John's stroke, Sheila found a speech and language therapist outside Cambridge whom they liked, and for three years John went regularly to two sessions on successive days, which allowed them to stay over in Cambridge and avoid two trips from London. Initially, the therapist had not wanted to take John: seeing him communicating so well without words, she did not think he would be sufficiently motivated to make the effort needed. She found that his ability to understand written words was normal, as was his reading and copying; his problems began, for instance, with matching a spoken word with its written equivalent. The therapist managed to teach John how to read for pleasure once more, but he wanted to speak, which as Sheila remarks was puzzling since he seemed to think his fluency was speech.[28]

After three years of therapy he had a vocabulary of a dozen words, a meager reward for the huge effort. Then, without warning, he stopped. "He stood in front of the fireplace, repeated his two utterances, hit the mantel-piece and said, 'Enough.' He was very decisive." Remember, his world had been one of scholarship and he could now read and listen, and this to an extent was enough. When Sheila challenged him about this, he proclaimed a few of his proper words before returning to his fluent word salad, with "arms open in his most histrionic manner, he said: 'Da woahs, da woahs, woahs, ach ga ga da woahs.' You see how eloquent I can be if you just leave me in peace."[29]

Sheila could not bring herself to force him to use his stumbling, impov-erished, new vocabulary. That would have been too hard on John and indeed on her. In a late, poignant paragraph she reveals her own collusion in his fluency:

> I would do anything to recall John's speech, except the one thing that might actu-ally work, which is to force his awareness of his voice by confronting him with it over and over on tape. . . . If we ridiculed or bullied him. . . . But I cannot sacrifice his voice. It's not that I cannot bear to hurt him. It's more a weakness in myself. I need his voice. I *do* understand it. It makes me laugh. It tells me things—about the world and myself and John's thoughts—that are beyond words.[30]

He was still John, and the man she loved was more vivid and alive in fluency than in his other stuttering forms of speech. She found his feelings and his presence were contained within his apparently effortless style.[31]

Hale quoted Howard and Hatfield that there may be 16,383 varieties of aphasia, reflecting the wide variety of problems with speech and language if one investigates in detail.[32] Yet John's loss of shared meaning, even when prosody and gesture remained—and he apparently had retained inner speech—laid bare the insufficiency of his speech. "To be dumb and yet have clearly formulated in one's own mind what one wishes to say, is one of the most grotesque predicaments which can be conceived."[33]

Those with Broca's-type aphasia struggling with words and yet often with clarity of thought and some understanding can become frustrated and angry, though eventually some sort of reconciliation is reached. Wint described how he learned to live more in his imagination.[34] John Hale had his reading and scholarship and, of course, could listen as well as perform. Those with fluent aphasia seem less troubled, perhaps because of their lack of self-monitoring as well as their continuing social performances.

We can still only peer from a vast distance into the inner worlds of those with aphasia in all its forms.[35] Where some recovery of meaning in words is possible, then we need to assist with this,[36] and if this is difficult, then we might try to help each person to become reconciled to their news way of living—whatever these may be. For her part, Sheila Hale was adamant that improvement might continue for years after stroke and that it was important to keep talking and keep including those with aphasia in conversation, where possible with the assistance of speech and language therapists. Social inclusion seemed less of a problem with John. The tragedy was that this depended on his default to a fluent aphasia with his single repeated utterance. His success revealed the importance of the surroundings to language in speech performance, prosody, and expression through gesture. His failures to communicate hard conceptual knowledge revealed the overwhelming necessity of this component of speech.

Whether one considers those with nonfluent aphasia, existing in a bardo-like world in parallel with those around them, sharing a lived space but devoid of words, or those with a fluent form whose vocalizations continue but with limited insight into the dissolution of their speech's meaning, not for nothing did Lam and Wodchis suggest that "Aphasia has the greatest negative impact on quality of life of any medical condition."[37]

16 Nothing in Mind: Abulia

The Russian neuropsychologist Lev Vygotsky, who died in 1934, spent his academic life, limited by tuberculosis to ten years, investigating the relations between thought and language, particularly in relation to children's development.[1] His background was as wide as it was deep, including the humanities as well as medicine; his thesis was on the psychology of art. It was a time when the behaviorism of Ivan Pavlov and John Watson was beginning to dominate, with measurements of responses taking precedence over other forms of investigation of the minds of others. In contrast, Vygotsky concerned himself with the inner lives and the inner speech of people. A pioneer in social neuroscience way before it emerged as a discipline, he considered culture to be an important way in which individual consciousness is built through relations with others.[2] His major work, *Thought and Language*, outlines his theory about the interrelations between speech, culture, and consciousness:[3]

> The primary function of speech, in both children and adults, is communication, social contact. . . . Egocentric speech occurs as the child transfers social, collaborative forms of behavior to the sphere of inner-personal psychic functions.[4]

He considered that speech, while beginning as social and in relation to affective/emotional experience and communication, became more cognitive and intellectual as the child realized that words carried external meaning and so could represent an object or an action. Thought could, of course, be prelinguistic and independent of language, but at some point it became verbalized, as speech also came to be "rational."[5] Just as Dunbar suggested the evolutionary origins of speech were social, so Vygotsky thought its ontogenetic development in children was too.

His other concern was what he called "the last 'why'":

Thought is not begotten by thought; it is engendered by motivation, i.e., by our desires and needs, our interests and emotions. Behind every thought there is an affective-volitional tendency, which holds the answer to the last "why" in the analysis of thinking. . . .

To understand another's speech, it is not sufficient to understand his words—we must understand his thought. But even this is not enough—we must know its [affective-volitional or emotional] motivation. No psychological analysis of an utterance is complete until that plane is reached.[6]

Running through his work there are two intersecting areas of thought: speech and its social and rational or intellectual aspects, and how the motivation for action and for those thoughts themselves is often beyond the intellectual and depends instead on the affective state or feelings of the person.

What Vygotsky did not consider was what might happen to speech and, in particular, to social speech or conversation if that affective motivation was lost.

Denise runs several nursery schools, while her husband James works as her PA. When I first visited their home, their son opened the door and led me to the lounge before calling James in from his office. He walked in normally and talked slowly but without any word-finding difficulty:

I'm fifty-seven . . . a company director . . . suffered basal ganglia damage . . . following a fit and it's left me without imagination. . . . I was never in hospital. I had a few fits before. This was two and a half years ago and after that my imagination was affected. I find it hard to think of things to say. . . . I used to post a lot online, funny stuff I came up with which I don't do any more. . . . I have no problems with words. I do Wordle every day and have all my memories fine, I can read and I can speak.

I can respond to stuff okay, but it's [problematic] coming up with new stuff.

Denise remembers what happened in more detail. After lockdown for COVID a group of their friends had arranged a party one afternoon. Knowing it would be "an almighty session," she decided to give it a miss. James, however, wanted to go—he had not such had an outing in two years—and she did not want to appear prudish by trying to dissuade him:

Things went wrong. He was brought home by the designated driver at six p.m. and seemed a bit drunk but nothing horrendous. I went to bed around ten p.m. and left him down in the sitting room. I woke in the early hours and found him slumped on the settee with a bottle of rum, so I left him there. When I came down in the morning he was on the floor. He got up and said, "I'm going up to bed," and disappeared.

The next day he slept, and Denise assumed it was because he had got "hammered." A few days later, he showed her bruising from his waist down. Denise suggested he go to his GP, not so much because of the bruising but because of how he had been behaving. He was all but silent and hardly responded to her at all. At the time she was furious with him: "I thought he was being a complete arsehole." He was adamant he did not need a GP, but she persisted and went with him, since she thought that otherwise James might not say anything.

The doctor put the bruising down to the way he had slumped on the settee or floor immovably for so long when drunk. He thought James's silence reflected depression, so he gave him a questionnaire and antidepressants. He was on the drugs for months, with Denise dragging him back to the GP several times to say they made no difference to his reduced speech or his lack of interest. The GP remained convinced James was depressed and just increased the antidepressants "to give them time to work." Denise never thought he was depressed, and neither did anyone who knew him; it was just that he hardly spoke nor did anything else much either, for months and months, except sit around. He could talk, watch TV, and could eat and drink okay, but he did not speak much and showed little or no interest in what was going on. Denise was getting nowhere, so she wrote a letter to the GP begging him to refer James to a specialist and sent their middle son, who is particularly tenacious, to read it out loud. In the end they paid privately to see a neurologist. An MRI scan was reported as normal, making them wonder if he was depressed after all. It was only two weeks later that the neurologist rang to say there was damage in a small area of the brain beneath the cerebral cortex known as the basal ganglia.

All this took a year since he was first affected, during which Denise, understandably, was frantic at the huge change in her husband and the lack of action from the GP. Before, she said,

> He used to write letters to the local paper, on cycling, safety issues, stuff about the local community and its problems and many other things. I used to call him Mr. Angry. That all stopped. He used to post a lot on social media; now he'll read it but not post anything.

His problem remained difficult to define: he had no problems with speech as long as he was asked something, but he volunteered nothing himself; his memory was unaltered; he had no weakness or reduced sensation; but there had been a more profound, almost unfathomable change in him.[7] His enthusiasm and passion for life were drained. Nothing seemed to matter, and he seemed no longer to want anything or to care about much, either. Denise had no way of explaining to others what she scarcely understood herself, even after a year of trying.

The neurologist said there was nothing to be done. When inspected carefully, James's MRI had shown small areas of damage in his caudate nuclei. But the neurologist did refer him to a research study trialing drugs for this condition. James went to the center and did lots of tests of IQ and linguistic and other skills, passing them all, and has been on the trial for a year or so.

I asked Denise how the doctors had explained such a difficult condition.

> I don't really feel that it has ever been explained to me properly. It is difficult to tell just how much was told to James as he doesn't relay the information back to me. Our GP hasn't spoken to us since we got the result from the neurologist. I learned pretty much everything I know about his condition from good old Google.[8]

The basal ganglia are major collections of neurons grouped together into collections or nuclei, called the caudate nucleus, the putamen, and the globus pallidus. These sit in the midline under the cerebral cortex and have wide connections between the motor and frontal cortices. They are thought to have a number of roles, including the assistance and planning of movements through connections to the motor cortex, and the selection and activation of programs for decision-making and thoughts via their connections to frontal and prefrontal cortex. Kailash, Bhatia, and Marsden studied the effects of damage to these.[9] In just over half of their sample, of 240 participants, various movement deficits were found. The commonest behavioral effect was a form of apathy, termed "abulia," which was seen in just over 10 percent.

Apathy can be defined as a lack of emotion, feeling, interest, or concern not due to a reduced conscious state, cognitive impairment, or emotional distress,[10] with a loss of initiative and spontaneous thought and emotional response. A similar condition can be seen in widespread atrophy or shrinkage of the cerebral cortex which is why much of this work has been done in relation to Alzheimer's and other causes of cognitive impairment.[11] It can also be seen in a number of conditions, including Parkinson's, Huntington's, and progressive supranuclear palsy, all of which involve the basal ganglia, and after overdoses of cocaine. It appears to depend on comparatively small areas of damage in basal ganglia structures, including the caudate nuclei and some of the thalamic nuclei.[12]

I asked James about his work. Since he had come in from their office where he was working online, fielding phone calls and emails, and managing their web material, I presumed all these parts of the job were fine. "I help [my] wife with nursery schools. I'm her PA." He had no problem with responding to phone calls when people ask for details such as prices or with looking things up online. Nor did he have difficulties with jobs around the house or at the schools, fixing a gutter, making arrangements with a plumber, or installing a new program on a laptop.

Did he enjoy the work? "Not really." There was an impressive set of bongos in a corner of the room. Did he enjoy playing those? "Not really, no." I wondered if work was better than doing nothing. "Sort of."

I turned to action and want. When I do something, I explained, I may look forward to doing it but I also needed to be able to start the action. Say, my shoelaces are loose though not disastrously so. I may want to do them up to make it more comfortable, and of course I know how to, but I also have to want to initiate the action. James had no problem with these. "I can get on with things." His motor memory and programs were intact and he could initiate action, but he lacked any inner sense of reward or enjoyment. Even while playing bongos, he said, "I am not really expressing when I play . . . it's more mechanics. You might enjoy those mechanics; I've been frog-marched into it. A friend badgered me into going to an open mike evening at a pub night once a month." I asked if he enjoyed going. "Not really." He would not go without his friend making him. On a coffee table in front of us was the newspaper, which he would read each day. "When sitting reading the paper, what comes to mind?" "Whatever is in the article."

Sitting in the sun, I wondered, does he enjoy that? "No." What if it was freezing cold? "I wouldn't enjoy that." Maybe because cold is uncomfortable, James agreed, but wouldn't he feel the opposite, enjoyment, in the sun? He was not convinced. It was as though his emotional experience, such as it was, might have rebalanced toward the negative. "Possibly."

He volunteered that he did enjoy eating and watching television, both comedy and documentaries. Some comedies made him laugh, though not as much as before, "A little bit," and then he was able to be in that moment and forget the condition. "I enjoy it because something is happening, feeding my thoughts, funny thoughts." Music might do the same, only less so. The comedy he liked included panel shows such as *QI*, *Mock the Week*, and

Have I Got News for You, all quite sophisticated programs full of puns, word-play, irony, and satire, and requiring a depth of knowledge and politics. He had not lost any knowledge or quickness of thought. He also watched the more intellectual quizzes, *University Challenge* and *Only Connect*, one a rapid general knowledge competition and the other a really difficult crossword puzzle/lateral thinking program. His intellect, memory, and language skills seemed intact.

I returned to his imagination, beginning with tasks and memories. Could he imagine walking to the pub? "Yes." Similarly, could he imagine stripping down his motorbike, what his best friend looks like, a Beatles song in his head, a sunrise, what it might be like in space, and what he might look like in ten years? Each time he replied, "Yes." But he could not contemplate writing a short story or a new tune on his banjo. It appeared that he could form pictures in his mind and think of some things he had not experienced but that new ideas were now beyond him. He dreamed as before but no longer daydreamed. Sometimes music would come into his head: "I find myself recalling tracks—usually that I've heard recently on the radio or adverts but sometimes completely random."

It was December, and I said how pretty their Christmas tree was.

> My wife chose it. I can respond but not start a conversation. It's not getting better or worse. I feel depressed over this. I am on a trial at the moment of a drug to see if that will help. They did . . . basic memory and cognition tests and a CT and MRI scans. That's about it. I did okay.

Denise explained how she managed his work:

> When James makes a work call, we make sure he knows exactly what to ask. At the moment we are talking to a loss adjuster for an insurance matter, and I have to tell him what to ask and to write it all down. We'd do this before his stroke but it is more important now. I tell him to remember to smile, or to ask the other person how she is.

She continued that he could work his way through problems but that now he makes more mistakes. It is as though he has less interest in his work as in everything. She asked if his interests in the business and the people had lessened. "Yes." She noted,

> He does not have the same desire to do things. He used to suggest I go to so and so while he'll work on the bike. He had a plan, whereas there's no plan now unless I come up with it, and even then getting the moment going can be slow. Maybe I don't give him interesting things to do.

I asked what he thought about when just sitting doing nothing. "I worry about things I have to do, fixing things." At other times his mind was empty.[13]

Denise continued:

> What I really want is for him to walk with me but he never did that before; he does not enjoy it. He loves food just not as much as he used to. Even those close to us, my parents, for instance, don't understand exactly what is wrong and that it won't get better. It is so difficult for people.
>
> The intonation of James's speech has also changed. For example, if I give him something to eat or drink, the way he says "Thank you" now sounds like he's talking to a waiter or someone he doesn't know rather than a family member. There is a loss of "warmth" in his speech as well as in his emotions. His voice sounds weaker in tone generally since his brain injury, and the loss of warmth and intonation gives the impression that James's mood is flat or even depressed (possibly reinforcing our GP's mistaken diagnosis). It's very hard for me to get used to as James was an exceptionally warm human being prior to this. I don't really know if he's lost his warmth or just can't show it anymore.

It had become clear that James's stroke had robbed him of much of the background emotional experience accompanying thoughts and actions. With that he had also lost his initiative to start a new task and lost his own spontaneity of action. It also involved a mental emptiness that he had mentioned when we first spoke: "nothing in mind."

What of his internal language and thoughts? I asked James if these were preserved as far as he knew. "Yes." But the problem is their expression through words? "Yes." I wondered about music: Is its enjoyment independent of language? "Yes." But to express appreciation, I might need words, but also embodied expression, gesture, movement . . . ? "Yes." Do you enjoy normally, but then its expression becomes blocked with words or expression? "Not really, no. Nothing comes to mind." It was not just the expression that was lost; his emotional experience was blunted.

Though sometimes there was nothing in his mind, at other times it was clear his mind was active with thoughts going round and round without any way of expressing them. I asked again about the main problem. James paused a long time, "Not knowing what to say." Some thoughts, I suggested, might be self-contained and well-thought-out and formulated into words, whereas others are not so clearly formed into words and sentences, such as appreciation of music or a nice day. Either way it must be so frustrating. "Very." He could not find a way of expressing his thoughts in words or within music during the jam sessions at the pub. At other times it was the thoughts that appeared to be a problem, maybe because even thoughts are rarely without some tinge of feeling. Later, I returned to this. What was so difficult? "Coming up with thoughts. I respond to stuff but I just don't come up with stuff spontaneously."

So, his problem might be less to do with tasks and actions and more to do with feelings? "Yes." What about mowing the lawn? I asked. You might need to do it; would you ever want or like to do it? I suggested we are motivated by reward, by feeling good, by either doing a job or afterward when it looks good. "I'll get some of those, I think." So where is the problem? I asked. Again, James thought long and deep, and then uttered one single word, "Chat."

"Chat"—informal, more friendly conversation rather than speech itself—was an almost insuperable and catastrophic loss. He could talk on the phone to a customer about their bill, to a builder face-to-face about their patio roof, but found conversational talk all but impossible, and this—above almost everything—was affecting him.

> I don't talk to my wife anymore. I cannot think of what to say. . . . It's affecting my relationship with my wife and children . . . with everybody. . . .
>
> Going for a meal, I could say where I would like to go, but there is not much conversation. Very dispiriting. I can respond but not introduce additional things. It's very hard on my wife and our relationship is destroyed really. That is hard. I cannot chat to her.

I suggested that since both knew what the problem was, might not that help? "Not really." I mentioned that not all expression was through language. "Yes, but . . ." I asked if it was easier to talk to family, friends, or strangers. "All the same."

By now it had been a long talk given his difficulties, and he asked, unbidden, "Would you like to talk to my wife again?" Denise had popped back to the office. I replied I would love to and reminded him that he had volunteered that idea himself. "Yeh." He dialed her, "Hello it's me. Do you want to come in?" She came in and sat down next to James.

> I feel weird talking about it in front of James, but it's nothing I have not said to him before. He was one of the most caring, helpful souls I have ever met. If someone had a problem, he was the first to go round to help. Now, even if someone is struggling in front of him, he no longer responds and I have to say, "James, can you help?"

It is so distressing for me and the family. I know if he could find a way to help, he would. I know that but it is so difficult. From the outside it may not be so apparent, but anyone who knew him before sees a different man. To people who did not know him before he may just seem reserved. He responds to a direct question fine but volunteers little or nothing.

I wondered if working might be better since he could then respond to questions normally. Denise agreed that this was now the case, though not initially: "Then it was very difficult and all the staff at the nurseries were concerned. It took a while. I would say, 'Don't forget to smile.'" I asked James if he felt anything when we smile. "Yes." His emotional responses, though blunted, were not extinguished.

Denise continued that James used to play instruments and loved to listen to others. He's now just begun to play percussion. He enjoys it but not as much. James echoed, "Yes, not as much." She continued,

I don't know. Our relationship has changed hugely. I don't feel as close to James as I used to. It's just so difficult to maintain . . . we were exceptionally close couple, together thirty-eight years; we worked together and played together.

Now I have to arrange things to keep myself sane without him. We tend not to do stuff together. Say, going out for a meal, or on holiday with just the two of us, would be difficult and even a bit pointless. We do things with family and friends but he tends to avoid these.

Being in a group was no easier:

Say, at a birthday party, people will go up to talk with him and get nothing back and so fade away from him and I have to fill in. He does not initiate much conversation at all. So, when you say something to him, you minimize it and receive a succinct answer.

Chatter makes people feel connected. Small talk has immense value in everyday life and indeed family life. Our middle son had his first child eight months ago. Though James is obviously happy about it, he doesn't say it, whereas before he would have been very vocal about it. It's not that he can't find the words; he cannot use them. A vocabulary test would be fine but having the lived experience of being a grandfather and expressing its pleasure is beyond him. Intellectually, he may be delighted, but since the emotional aspect is no longer there, the whole experience is reduced and he's unable to express. He said that he does not have any imagination anymore. It affects everything, very much.

I asked again if he could imagine how others are feeling, and whether he could still empathize with them. "Yes." Can you imagine what I might be thinking and feeling to meet you? "Yes." But you cannot say it. "Maybe."

I wondered if he could inquire how someone else is feeling.

I did the other day. Our investment guy I phoned up to postpone an appointment and I found out that his wife was being treated for cancer and I managed to ask him how she was.

This was the first time he had managed such a question himself since the episode. Denise mentioned this independently, too: "I heard you say you hoped she was okay and you sounded concerned about her, whereas before you might not have sounded concerned or known what to say." "Yes." Denise was surprised but delighted. I suggested that forming the idea as a response and then turning that into words in real time might be difficult. He agreed.

I asked what the stumbling block was. "Having things to chat about." The usual things—weather, work, children, gossip—were out of reach. "Yes." And if others asked him about these, he could not answer. Conversation was so much more difficult than being a PA and answering the phones. "Nothing comes to mind. It is difficult." I asked if it was easier if people were patient and waited for his answer. "No."

I wondered if most of the time, when nothing is happening and he is empty, did he feel empty, or was he aware of the emptiness? "Just empty." When he was being activated he was aware of that, but when "empty" he was shut down. "Yes."

On another occasion I asked if he felt it was getting better. "No." Can you accept it or is it frustrating? "I accept. Always have, pretty much." I mentioned a family with an adolescent boy with Möbius and impoverished emotional experience. The family talked about what they were feeling in words the whole time, hoping that might allow the boy to experience and express emotions more too. I wondered if that might help him. "I don't think so." "If I show happiness, are you happy?" "Yeh." "If I laugh, you'd laugh and be happy?" "Yes." But he did not think verbalizing emotions would help.

Denise mentioned that if they had people round she could sense him relax when they leave because he could not join in. They used to play and sing with a group of friends at home. Recently, they came around again and she was completely amazed that James could sing all the songs. A few days earlier, she had heard him sitting on his own, singing to his favorite music. She said, "I burst into tears. It was joyful." It was the first time he had expressed such joy since the stroke. I asked if his friends came round for a sing-along, loud and all together, might he enjoy that? "Yes, I think I would."

I mentioned that I had been in Sweden and on my own for three months. In the evening I'd go to a café or bar and, despite not understanding anything that was spoken, I'd enjoy being in the presence of other people. Might he feel the same in a chatty bar or at home with friends even if he contributed little? "Not really." He often felt anxious about his lack of reply and relaxed when they left.

Denise wondered if practicing role play beforehand might help, if he could be motivated to practice. He was not convinced.

> I went to speech and language therapy and they suggested a list of questions I could ask. But, you know, it's not conversational . . . it's mechanical. I could go through the list but it would be repetitive. It's the initiation [that's difficult].

Denise mentioned that the nursery staff at her nursery have to act every day to an extent, putting aside their stuff to be cheerful and full of energy. "James, you may have to pretend if you can." But, I thought, there were problems with the energy required and with the uncertainties of social interaction. James might smile and ask how a person was, but that person might have just won the lottery or their cat may have died and James might not be able to respond appropriately. He agreed this was difficult. Denise returned to his work and the time recently when he had installed something on a laptop:

> The other day I asked James to set up a new email account on a laptop at one of our nurseries. He can do that easily, standing on his head. He went out, put it on, and came back without telling anyone. The manager asked me if she could use the laptop. She then phoned me to say she realized now why he had popped in. He had not thought to tell her.
>
> Now with orders, I ask him to do the order and then let them know.

Apathy, as we have learned, may be defined as an absence of emotion, feeling, interest, or concern or a lack of motivation not due to a reduced conscious state. In their review focusing on apathy associated with disease of the basal ganglia, Levy and Czernecki criticize the above definition, preferring to concern themselves with measurable behavioral outcomes, rather than resort to speculation about internal states, echoing behaviorists:

> Motivation is a psychological concept, . . . with several different theories . . . which is difficult to transfer to a physio-pathological basis. Defining apathy as

lack of motivation is an inference, and apathy may also be the clinical expression of different underlying mechanisms.[14]

Instead, they suggested apathy be seen as a reduction in actions, which can be measured in terms of self-generated, goal-directed behavior. Levy and Czernecki were not blind to how these reductions might be the result of differing problems in the brain systems underpinning these and identified three different underlying problems that may affect these behaviors.

There may be a difficulty in processing the emotions associated with an action, either because of a lack of or reduction on the will to perform or maintain it or because of a disconnect between doing it and its future consequences. The person may no longer have the working memory to perform an action or find it difficult to produce the subtle new ways of moving that some tasks require. The last and most severe problem, they suggest, is in summoning the thought or initiating the motor programs within the brain in order to start and finish a behavior, which they call "autoactivation." With this, they thought, came not only difficulties in initiating actions or thoughts, but also an internal mental emptiness, even when their externally driven actions were relatively preserved.

James's problems were manifest in some behavior but reached way beyond behavior alone. They related to a mental emptiness and a profound reduction in the experience as well as the expression of emotion. His apathy has the most profound effects on his ability to self-motivate, to feel, to empathize, to engage, and, hence, on how he views will and even self. Yet cutting across these is the striking difference between talk involving externally driven factors about doing things, like building work or the laptop, and his profound loss of small talk, or just "chatting." The deep relation between Vygotsky's two paramount ideas—that a primary purpose of speech is for social communication and that in the final analysis motivation was emotionally driven—were revealed with a striking starkness in James's new situation. Despite his severe losses of emotional experience and emotion, James was still able to respond to questions and retained movement, memory, and speech, but was almost paralyzed in that most everyday and yet profound way of interacting with his wife, family, and others—small talk—with huge consequences for his social life and indeed his own sense of identity. Denise related,

> I find it so difficult because James does not express what he feels. Outwardly, it appears it has not mattered but of course it matters; it's not being able to share

with others. Relationships have gone. Some of James's family do not even know and he says they do not need to know since they do not live in the country. But if they ever visit [they] will wonder what has happened. Our sons may like to read anything you write, since they do not really understand. They are so distressed they almost pretend nothing has happened. They would come round and start talking to me and then sometime later might notice James and say, "Oh, hi Dad."

I asked James if that upset him. "No." Denise continued,

I felt really upset after talking to you last time. I felt I had built a shell around the awful upset feeling inside and every so often it cracks and floods out. You try not to think about it all the time.

She turned to James, and asked if he felt upset when they talk about it. "A bit."

To express yourself as a thinking, feeling being, language seemed crucial, not only the words and their prosody and gesture, but the emotion all these carried and perhaps above these the emotional motivation itself, as Vygotsky had written. I mentioned the song "Julia," which John Lennon wrote to his dead mother Julia: "Half of what I say is meaningless, but I say it just to reach you."[15] Denise agreed with the sentiment. I suggested that for James it was important for people to know that though he may have reduced expression, he, as James, a person beyond that and within that, was still present. Denise: "Absolutely." James: "Yes." It sounded a little like a personal prison. James just nodded and said, "Yes."

Denise's sorrow and frustration were clear, but for James the reduction in emotional experience may have blunted his regret. In his account of his own descent into tetraplegia after operations for a spinal tumor, Robert Murphy wrote of the alterations that occurred in his will, even to make a relatively simple movement:

For a while I tried to will the legs to move, but each futile attempt was psychologically devastating. . . . I was saved from the edge of breakdown because the slow process of paralysis of my limbs was paralleled by a progressive atrophy of the need and impulse for physical activity. I was losing the will to move.[16]

James's caudate strokes may have not only reduced his will and abilities to feel and express emotion, but also diminished his frustration at this loss.

I hoped that he might feel some recompense in family and friends knowing of his condition and knowing that James was still there inside. James agreed. After all, I said, we all knew friends and family and work colleagues who were curmudgeonly and miserable but still loved and indeed particularly loved for being miserable in their own way. I tried to reassure him that his new condition did not preclude a human dimension nor friendship nor love. There is something beyond expression that remains, especially once his difficulties were understood. James just said his simple, "Yes."

What is so difficult to convey was that as Denise and I talked, James was the third person in the conversation. Though his utterances were minimal, his engagement and presence were not. Denise was pleased that he wanted to explore his situation. She had not known why, but that he had wanted to—actually *wanted* to, seemed important. She had found it useful, even though it had been hard to revisit or explore for the first time. A draft had reduced her to tears. "It never goes away. Just reading about him, putting it into words and talking about it . . . to be understood . . ." she tailed off. Yet she was adamant that both she and James wanted their story told, and both were happy with the result.

Trying to explain James's situation was difficult enough, as conversation proceeded by me asking questions, to which he would give one-word or one-phrase answers. I did not want to lead or, more accurately, express things that he did not accept, but it also required me to try to imagine an interior world with such a breakdown in emotional experience and its expression. As I left, I said I hoped it had not been too difficult for James nor too anxiety-provoking. He smiled, faintly, and said no to both. Small talk was hard talk, but deep talk was different.

17 Conclusions: Residing in Voice

John Hull, a professor of religious education, was perhaps most famous for writing about the experience of going blind in his forties. He told me that, knowing he would go blind, due to a rare, untreatable condition, he had consciously developed visual representations of his loved ones to remember them when the day came. He became depressed not on becoming blind but sometime later, when these visual memories faded. So visually dominated was his memory for others that he had lost the identity of his family and friends. Over the succeeding few years, and with considerable effort, he had been able to find that slowly sound took over and that he imagined fully, wholly, delightfully, a person's mood and identity from their voice, words, and volume, pitch, modulation, and prosody; all was in the voice. Even his judgment on new people agreed with his wife's, his from voice and hers from voice and sight.[1] For him, he said, people now "resided in their voice."

Hull continued that for him to know anything, the person had to speak. In silence he was lost. Speech is, perhaps, even more than gesture and facial expression, an act, something we do, and has elements of performance. This is revealed in the way in which people with cleft were sometimes reluctant to speak since it brought attention to them. In expression we also can become vulnerable and exposed. Jo, with Möbius, always performs and looks for feedback from the other person to gauge their success. As Carol Gilligan wrote, "Speaking depends on listening and being heard; it is an intensely relational act." Voice is "something like what people mean when they speak of the core of the self."[2]

I initially thought it might be possible to encompass all problems with speech, not realizing how widespread and frequent they are. As a result, many conditions are missing from this work. The community of those who

have had a laryngectomy (removal of their voice box) is one such example. Having discussed speech in terms of both its language and its prosody,
meeting with those with an impairment regarding the prosodic elements
of the spoken word would have been of interest, to explore some of the
consequences of the absence of the nonlinguistic, emotionally expressive elements of speech. There is a rare neurological condition, aprosody,
that parallels aphasia. Whereas aphasia is seen in dominant hemisphere
problems, aprosody has a nondominant, so usually right, hemispheric pre-
dominance.[3] It was considered within a short essay by Oliver Sacks, "The
President's Speech":[4]

> Natural speech does not consist of words alone. . . . It consists of utterance—an
> uttering-forth of one's whole meaning with one's whole being. . . . For though
> the words might convey nothing . . . spoken language is normally suffused with
> "tone," embedded in an expressiveness which transcends the verbal.

A group of people with different types of aphasia who are listening to
the speech of a president (most likely Ronald Reagan) can discern its false-
hoods despite not understanding his words. People with aphasia, Sacks suggests, may lack language but become more sensitive to the feeling tone that
accompanies language through prosody and gesture: "one cannot lie to an
aphasiac." They recognize "the grimaces, the histrionisms, false gestures
and, above all, the false tones and cadences" and saw through the president's
speech, even though he had been an actor. Sacks then describes a teacher
who had developed aprosody and was unable to discern anything in a person's speech beyond the words employed. Initially, naturally, she paid more
attention to the person's facial expression, but then her sight failed too, leaving her with their words alone. She, too, saw through the president: "He is
not cogent . . . either he is brain damaged, or he has something to conceal."

In another context, Jonty Claypole comments that

> people with speech disorders have always been well placed to expose the glib-
> ness, deceitfulness and intolerance of other modes of communication inherent to
> hyper-fluency. They do so because they have a fundamentally different relation-
> ship to language than those who consider themselves fluent. It is still a tool for
> communication, but one which is inherently unreliable and requires constant
> vigilance to avoid tripping them up. By looking at their own speech in this way,
> they become aware of the ways others use and misuse language.[5]

Missing too from my account is any consideration of the slow losses in
language and speech that can accompany dementia. I approached several

physicians, psychologists, and patient support groups without success. Nicci Gerrard has considered some of the consequences of losses in speech in dementia, following the death of her father from the condition.[6] She suggests that in the early stages, people with dementia may be able to express their feelings, and then she wonders, as this becomes more and more difficult and then at last barely possible, what that must be like. She quotes Carol Gilligan: "To have a voice is to be human. To have something to say is to be a person."[7] In one form of dementia, frontotemporal, semantic memory is most affected, so comprehension of speech drains away first.

In one care home,

> a tall skinny snaggle toothed woman came up to me, clearly upset and trying to tell me something. But I couldn't make out a syllable; she was talking in nobody's language. Words that weren't words poured from her.[8]

Alas, even Gerrard, a humane, passionate, and eloquent writer, was unable to reach those with dementia as their speech and expression failed. Oliver Sacks once said to me that he would like to write a book about the subject, "a reverse Piaget," after the pioneer of studying children's development, but it was not to be.

At a more practical level, the Alzheimer's Society of the United Kingdom has a useful website with advice about communicating with those with deficits in speech production and perception.[9] They offer what may appear to be obvious but is enlightened common sense. It is important that each person is comfortable and relaxed at the start and that they are in a quiet space. They suggest having something suitable for conversation and maybe to think of the last time you spoke. Eye contact and attention to body language is recommended, as is slow, calm speech and, where appropriate, enjoying laugher with the person. As often, this advice for communication with those with a problem is sound advice for anyone.

One risk factor for or association with dementia is hearing loss in middle age and beyond.[10] Uchida et al. discuss the possible mechanisms. The *cognitive load hypothesis* suggests that if hearing becomes difficult, more cognitive resources are given over to this and other domains become overloaded. In the *common cause hypothesis*, cognitive decline and hearing loss are presumed to have a singular independent cause. In contrast, the *cascade hypothesis* postulates that hearing loss may accelerate cognitive decline, possibly because it comes hand-in-hand with social isolation, loneliness, and then depression, which can lead to impairments in cognition. Evidence for the importance of any one of these is not firm, but there is an increasing recognition of the significance of social interaction for all with dementia, and crucial to this is conversation with others.

A further absence from these accounts is stuttering, or stammering.[11] Jonty Claypole, himself a stutterer, has written a rich account of the subject.[12] He begins with *The King's Speech* (2010), a film about King George VI's profound stutter, which was allegedly helped by an Australian speech therapist, Lionel Logue. Claypole questions the effectiveness of the therapy and poses the question, "Why does it matter that the king should be fluent?"

> His problems with communication were exacerbated not so much when he stuttered but when he tried not to. If he had stuttered, he might have had more humanity and cadences might have matched the meaning better. . . . His speeches were exercises in simulating fluency. . . . Fluency and good communication often go together but are not the same thing.[13]

He mentions that more than one million people in the United Kingdom are nonfluent and have to contend with bullying, avoidance, and attempts at concealment, 5 percent of children and 1 percent of adults. Rather than being the norm, those fluent in public speaking are "a charismatic elite."[14] Many of us are suspicious of good public orators and professional charmers and beware of politicians as purveyors of snake oil. Sometimes, however, the reverse may be the case. The post–World War II UK Labour politician Aneurin Bevan, architect of the welfare state, stuttered and would use it as a tool. Claypole wonders whether Bevan was effective despite or because of this. How did he overcome his stutter? "By torturing his audience."[15] Joe Biden, it was suggested in 2021, lived with his mumbling diction being put down as cognitive decline as a price for concealing his stutter.[16] In fact, Biden has written about the debt he feels to his speech problem: stuttering was "the best thing that ever happened to him": "It gave me insight into other people's pain. I am sure it has made me a more empathetic person because I know how much lies beneath the surface of words we say. I am forever listening for the emotional subtext."[17]

Claypole lists stutterers, including Lewis Carroll, Henry James, Christy Brown, Wittgenstein, Stephen Hawking, and Marilyn Monroe. James's stutter led to him to use long-winded circumlocutions, while Monroe's breathy, sexy speech reflected her way around her stutter.[18] Claypole quotes

Somerset Maugham that the first thing people should know about him was not his homosexuality, which some suggest is the cause for the sense of loneliness in his work, but his stutter.[19]

Musicians have been drawn to song to reduce their stutter. B. B. King found an outlet in singing: "Words aren't my friends. Music is." John Lee Hooker stuttered when he spoke but not when he sang.[20] Stutterers have been used to comic effect, of course, and with some empathy in *A Fish Called Wanda* by Michael Palin[21] and in a UK TV series, *Open All Hours*, by Ronnie Barker. Few uses were angrier or more defiant than the stutter adopted by Roger Daltrey, in The Who's song "My Generation," with its legendary line, "Why don't you all f, f, f, fade away."

Claypole relates the story of his own speech therapy, both as a child and later, in his thirties. The key aspects were the realization that stuttering was a neurological motor problem and not a psychological one (without diminishing the psychological issues that might result from it) and that the best way to manage it was to speak to people without breaking eye contact and without implicit apology. He quotes Willie Botterill, his therapist:

> When we started out, it was simple, our job was to get you fluent. Now it's different. How much you stammer is not particularly important. . . . If you can get to the point where the amount of fluency is not the issue and communication becomes the important bit, then that substantially alters both the way the individual and the world around them manages and thinks about it.[22]

The Irish raconteur Patrick Campbell used his to great effect, too, throughout his career: "Most empowering is seeing stuttering as my voice and part of who I am rather than something to be fixed."[23] His stutter had a rare comic timing all of its own.

Claypole concludes by asking how one can change the way society thinks about those with speech difficulties. One problem is that these are so diverse and different that they have little coordination or community. Though there are societies supporting those with dyslexia, stammering, and aphasia, as there are for those with neurological causes of dysarthria, through organizations for stroke, motor neuron disease, and Parkinson's, there is no one organization offering support and advocacy for those with speech impairments. Claypole quotes Erving Goffman: "The peculiarity of speech disorders discourages any group formation."[24] One reason may be that speech impairments are sometimes not considered a predominant problem compared with, say, tremor or reduced mobility in Parkinson's, or that the condition itself is rare, as in Möbius or dysphonia. Claypole suggests a group for those with "communication diversity."[25]

A thread throughout this account has been the way in which difficulties with speech affect a person's identity, whether at work or socially, and whether with partners, children, or friends. We began this book by citing Dunbar's theory about the origins of language as being a way to socially groom a larger number of people than might be reached by other (nonhuman) primates' tactile grooming behavior. Though stressing the social origins of utterance, the theory can be applied only so far. Remember how Jo was appalled by Dunbar's theory, always tailoring their words and use of prosody and embodied expression, reduced though it was by their absence of facial animation, to each person and each group they met, and how Jo constantly monitored their reactions. Jo's social grooming remained personal and performative.

In addition, though utterance may have allowed emotional affective communication with a larger social group, the evolution of speech and language enabled the development and sharing of intellectual concepts and ideas previously in unimagined ways, expanding our ability to remember, predict, and plan. Speech then can be both an affective emotional communication, seated in embodied and vocal emotional expressiveness through prosody and gesture, and a medium for communication of ideas and thoughts.[26] The present account has focused on the social consequences of problems with speech since these have been foregrounded in people's experience of difficulties with speech.

A curious and unexpected finding was that so many people with differing conditions thought their voices sounded "normal." The speech of some with cleft was all but incomprehensible, even though they had little awareness of it, while several with cerebral palsy sounded normal in their heads despite their speech being markedly affected. Those with fluent aphasia also appear to think they are speaking normally, though in this case it is usually explained by a stroke-related deficit in the self-monitoring of speech, and indeed that was part of Wernicke's original hypothesis.[27] One might postulate that those with congenital conditions like cerebral palsy and cleft have heard no other voice and so may presume that their natural voice is indeed normal, like others. This raises the question of how we hear our own voice. Of course, it never sounds the same to us as to others. This is at least in part because not only do we hear it through the ears, as others do, but we also receive the sounds as they are generated through the bone and flesh of our heads, whereas what others hear is through the ears alone. This may not explain all.

Speech is a rapid, ultra-skilled performance, perhaps the most skilled and fast of all our movements. We speak without being conscious of how the brain, nerves, muscles, and vocal tract coordinate to produce speech. It requires fast central motor programs in the brain—it is these that are clogged in spasmodic dysphonia. So rapid is speech that peripheral feedback of the result is too slow. There is evidence that for such actions the brain instead uses signals derived from the outgoing motor commands, so-called efference copies, which are then compared with the forward predictions of those commands to detect errors. These models, derived from engineering and applied mainly to control of limb movements, may also be relevant to speech.[28] What is not clear is that when people say they hear their inner voices as being their normal, do they have any unconscious access to these models on which to base this?[29] Whatever the case, the illusion of the normality of speech is often deeply entrenched in people with these conditions.

In other acquired conditions, like Parkinson's, people may have such a slow decline in the volume and clarity of their speech that they are unaware of change in their vocal performance. Others may hear a change but presume the person with Parkinson's is becoming duller and less interesting rather than exhibiting a sign of the condition. One woman with Parkinson's described how others saw her changed identity. She belongs to an embroidery group whose teacher once said to her, patronizingly, "How is our little Mary, sitting so quietly in the corner and getting on with her work?" In her youth and even a few years before, "Mary in the Corner" would have had a sharp response. "I was never the one who sits in corners and quietly gets on with things." But now, sitting quietly with her embroidery or art is to be in a safe space after several years of her condition, a haven from the awkwardness of conversation and safe in the healing proximity to her embroidery friends.

One of the more intriguing aspects of speech problems is the way in which these can lead to alterations in thought and its expression. Thought, language, and speech usually are just one seamless whole. In contrast, Minae, with cerebral palsy, has to tailor thoughts to what is possible to say, with her ability to pronounce paring back her expression and, at times, her thoughts. Aphasia is difficult to understand in this regard, but Faye

has found her speech does not always correspond with what she is thinking, while John Hale's cognition was largely intact when his utterances
were minimal. The inner experiences of others with the condition remain
beyond us.[30] Remember too Jo, with Möbius, always composing a sentence
in their mind before saying it, always thinking before expressing, feeling
feelings but having to think them too in order to express them.[31]

Jonty Claypole mentioned a similar experience reported by Betony Kelly,
a stutterer who wrote in her blog:

> [the] seamless connection between your thoughts and your mouth just isn't there.

Claypole continued that

> As a result a disfluent speaker is never quite in the moment. However present they
> appear, an inner process is at work: scanning ahead, determining when to engage
> and how; sifting words to remove the clunky grit.' . . . The disfluent speaker has a
> fundamentally different relationship with language to fluent speakers. It is some-
> thing which cannot quite be trusted . . . a matter of wrestling with an unfamiliar
> tool . . . because language is what binds human society together. No wonder then
> that feelings of alienation, of simultaneously observing and participating in any
> social situation, of never quite being in the moment of verbal communication are
> so frequently described. . . . There is the isolation . . . of feeling of detachment not
> only from others and from language, but from oneself.[32]

Later he describes how this detachment was increased by the

> very nature of a speech disorder. Given that speech is a tool for communication,
> then anything which hampers its efficacy separates an individual, even just a
> little, from usual comforts of human discourse. . . . We use speech to describe our
> thoughts and feelings. When this is faculty is disordered, we find ourselves at one
> remove from the thoughts in our mind.[33]

Now think of the worlds of those with severe cerebral palsy whose speech is
minimal or absent, and each sentence is measured in minutes.

Within a small series of narratives, sampling a wide variety of different
conditions, it is also difficult to achieve a balance between the attitudes of
those portrayed. It is clear, however, that for many with, say, vocal cord
palsy, spasmodic dysphonia, or cerebral palsy, problems with speech were
viewed negatively and at times led to disastrous consequences. But impairments are not always thus. Jo, with Möbius, as an adult embraced and
enjoyed life perhaps more than others, while after his stroke, Bill had found
a peace that previously had eluded him. In their accounts of living with
aphasia, Greenfield and Ganzfried found that many experience gratitude,

explaining how "it has made them wiser, stronger and more humane."[34] Certainly, some people do find some positive consequences within their condition,[35] but in this sample, they appear to be in the minority.

Differences in speech and especially conversation become clearer in those for whom these are problematic. Lizzie, with Parkinson's, could get by as a teacher, but found gossip and social chatter difficult. The rules of conversation at work may be easier and less fluid than during social conversation, where ebb and flow is the norm and unexpected turns are common.[36] This was seen most starkly in James. Severe abulia allowed workplace conversation while abolishing small talk. Social chat was revealed as being completely dependent on feelings and emotional motivation and thus different from more rule-based work interactions.

How, then, to help those with speech problems? This account began with problems in the vocal cords and cleft, and in these cases, where there are surgical or medical methods to alleviate them, they should be pursued. At times it seems that diagnosis is hampered, say, in spasmodic dysphonia, by unfamiliarity, thus better training of specialists would help. But not all conditions are so treatable, and even those that can be improved are rarely removed completely. Continuing support and assistance can be provided by speech and language therapy, as we have seen. In those with aphasia, Sheila Hale and Bill both said that the important thing was to continue to include people with aphasia and talk with them, and keep talking, for improvement may occur far later after a stroke than one imagines. Alex Leff suggested that each person with a speech condition should be given a minimum of five minutes to talk about their day without interruptions or completions of their sentences by a friend or family member.

Leff and coworkers also have clinical evidence of how intensive speech therapy can improve outcomes after post-stroke aphasia in a cost-effective manner.[37] Most effective is early speech and language therapy, for at least twenty to fifty hours. Unsurprisingly, the best outcomes are in those under fifty-five and with less severe aphasia.[38] Their success in improving outcomes also shows the effectiveness of speech therapy and engagement in those with aphasia or other speech impairment. Whether you "say it loud" in a group of people with Parkinson's, meet others with cerebral palsy to compare notes, or are encouraged to chat after aphasia, use of the voice and the conversational, shared, social space seems the way forward, since without that one can become isolated and lose rather than improve recovery and linguistic skills.

James Partridge had severe burns to the face and body as a student, and subsequently became hugely influential in the community with facial disfigurement.[39] One of his many insights was that though those with disfigurement will be accepted by the family and friends, they will still meet new people every day who stare or question. People with a disfigurement need the adroitness to approach and disarm others. Partridge developed social skills courses to enable those with disfigurement to manage other people. Though this must be difficult for those with problems in speech, finding ways of putting new people at ease might also be explored. Partridge also sought to educate employers and society more widely on how to accept and look beyond disfigurement.

From Partridge's example, we should treat medically and surgically and through speech and language therapy what can be treated; understand in depth the problems that arise from speech being hard; improve society to enable and facilitate full engagement within it by those with communication difficulties; and accept, support, make allowances for, and celebrate diversity. We will all surely be richer for that.

The rest of us have one other task: to listen and to be prepared to give more time for those who need it. George Eliot needed not only to utter in order to be, but she also needed—and presumed—someone to listen and indeed with whom to converse.

Medicine tells of its latest advance, and while treatment and cure are more than satisfying, one of the other rewarding aspects of the work is simply to meet with people and listen to their problems. In the clinic I sit down with the patient and ask them what the matter is. They often say, "The doctor says I have such and such." I ask again, less interested in someone else's diagnosis than in them and their problem in their own words. Some have already seen several doctors but never felt listened to and can understandably feel aggrieved. I give them time to talk themselves out. Not only will they usually be gratified that someone has listened, but in talking they usually explain their problem in such a way that the diagnosis may become clear. Oliver Sacks once said to me, "There is only one rule. Always listen to the patient."

In her recent book, Kate Murphy, a journalist and contributor to the *New York Times*, delves into the importance of listening. As she suggests, "to listen well is to figure out what's on someone else's mind and demonstrate that you care enough to want to know. It's what we all crave; to be understood."[40] Listening is not passive but reflects interest in another. She quotes Studs Terkel, the master of oral history: "The obvious tool of my trade is the tape recorder. But I suppose the real tool is curiosity."[41] A review of one of my earlier books praised my "Proust-like emphasis on the specific, and Magic Realism way in which a small segment of life is drawn in exquisite detail."[42] Then came the killer criticism that of course I had only talked with extraordinary subjects and that the life of many others is not like that. I thought that this was maybe the greatest compliment of all—unintentional though it was—for as Murphy writes, "Everyone is interesting if you ask the right question."[43] She also quotes the writer Elizabeth Strout: "I have listened all my life. I just listen and listen and listen. . . . If you do listen, you can get an awful lot of information."[44]

One other aspect of listening in this way is that it comes without judgment. The US writer George Saunders wrote on Chekhov, who in addition to being a writer was also a physician:[45]

> What I admire most about Chekhov is how free of agenda he seems on the page—interested in everything but not wedded to any fixed system of belief. . . . He was a doctor and his approach to fiction feels lovingly diagnostic. Walking into the examination room, finding Life sitting there, he seems to say, "Wonderful, let's see what's going on."[46]

Wittgenstein remarked how difficult it can be to listen without prejudice, but this is what medicine teaches and requires, and is one of its joys. To spend several years seeking out those with problems and talking and asking about their experiences, either in person or online via Zoom, has been a fascinating privilege and pleasure. To go to people to explore their experiences in their own words can never become ordinary. The eminent British writer Ian McEwan wrote, "Imagining what it is like to be someone other than yourself is at the core of our humanity. It is the essence of compassion and the beginning of morality."[47] Without criticizing such a great novelist, my parallel method is to ask and to listen.

Looking back, one of the more extraordinary aspects has been that I have never found it difficult to hear what a person has been saying, apart from some with cerebral palsy and with aphasia. Despite their sometimes severe experiences of speech impairment and its consequences for work and social life, when conversing one-to-one in a quiet space and with sufficient time, I have never found listening and conversation a problem. Sometimes a person's speech might have been low in volume or require more breaths, but tuning in was possible. Sometimes, say, in those with cerebral palsy, a personal assistant was required, but this was rare. What was more frequent was that listening and chatting took place with different timing and duration. Just as asthmatic children and adults may speak rapidly as their breath allows, so those with speech problems may be slower and need more time. In her book on disability and politics, Alison Kafer mentions the malleability of time and how many people with various conditions are slower, whether to travel or dress, and this needs to be accepted.[48] There is a rhythm to conversation, and if someone has a problem with speech and is slower, then so be it. Remember Jeppe's family waiting ten minutes or so to hear one of his jokes. Similarly, if the volume of speech cannot be sustained, then the listener's job is to compensate if possible. It is salutary to reflect that many people whose experiences are narrated in this book have had a problem not because of how they speak, but because their speech requires a little more care on behalf of those listening.

Speech is something we *do*; we talk of speech acts with an element of presentation and display. Minae reflected: "Speech matters to me, it is a kind of performativity." It requires a person to put themself in relation to another even when this may be difficult, and, as someone speaks to us, so we should listen, and if their speech is difficult, listen better. Conversation is social and carries an ethical dimension. *Hard Talk* is about reaching out to others through voice and how this, in turn, is one way in which we are defined, whether we are speaker *or* listener, and how this reduces the distance between ourselves and others. Speech does matter, whether it is effortless or onerous. A recent newspaper article reported on a small, remote Sardinian village with eight centenarians.[49] Their secret? "If you talk, you live well." Everyone needs to talk, and should be allowed and enabled to do so.

Postscript: Inner Voices in Aphasia

Vygotsky was one of the first who considered not only the social origins of speech but how children's speech evolved as they developed, from the first utterances to becoming more reflective and initially thinking aloud, to then talking to oneself, and then to internal talking and thinking.[1] Subsequently, these have been defined as "inner speech" and "private speech."[2]

"Inner speech" remains, unsurprisingly, difficult to pin down. Vygotsky thought there were two intersecting and overlapping circles, difficult to tease apart. "The meaning of a word represents such a close amalgam of thought and language that it is hard to tell whether it is a phenomenon of speech or thought. A word is speech, a word is a generalization or a concept."[3] He was equally clear that there are vast areas of thought without relation to speech.[4] "Speech and thinking are not related. There is a pre-linguistic period of thought (in children) and a pre-intellectual period in speech." Fernyhough has suggested two distinct forms: "expanded," which retains many of the phonological properties and turn-taking qualities of external dialogue, and "condensed," with semantic and syntactic transformations that take it closer to "thinking in pure meanings," described by Vygotsky,[5] and to Pinker's mentalese (described below). The former might lead to activation of language and speech areas of the brain more than the latter, which might activate default mode networks. However, Alderson-Day and Fernyhough were skeptical that inner speech was a "language of thought," not least because it is "largely untestable and often conceptually muddled."[6]

Despite such doubts, they go on to discuss what might be the functions of inner speech. At one rather mechanistic level, it may act as an offline abstract code, a tool for internal planning, rehearsal, or rumination, and may have evolved under this selective pressure. They also quote Vygotsky (and Luria)—approvingly—as viewing inner speech as an important, even revelatory evolutionary development, which "wired together" independent neural systems through social experience. The basic tools necessary for this progression—such as a phonological loop, the sound-related part of working memory, and the capacity for verbal rehearsal—may already be in place relatively early in childhood to assist speech production and language learning. Subsequently, with social interaction and communication, these tools may support cognition in increasingly complex ways. It is for this reason, they suggest, that inner speech is used much more often as a synonym for thinking than it is, say, for auditory imagery, a "motor-based linguistic tool that has by chance created an inner life."[7]

The importance of inner speech was also stressed by Oliver Sacks in his exploration of sign language and deafness:

> Dialogue launches language, the mind, but once it is launched, we develop a new power, "inner speech," and it is this that is indispensable for our further development, our thinking. "Inner speech," says Vygotsky, "is speech almost without words . . . it is not the interior aspect of external speech, it is a function in itself. . . . While in external speech thought is embodied in words, in inner speech words die as they bring forth thought. Inner speech is to a large extent thinking in pure meanings." We start with dialogue, with language that is external and social, but then to think, to become ourselves, we have to move to a monologue, to inner speech. Inner speech is essentially solitary, and it is profoundly mysterious, as unknown to science, Vygotsky writes, as "the other side of the moon." . . . Our real language, our real identity, lies in inner speech, in that ceaseless stream and generation of meaning that constitutes the individual mind. It is through inner speech that the child develops his [sic] own concepts and meanings. It is through inner speech that he achieves his own identity; it is through inner speech, finally, that he constructs his own world.[8]

Passingham echoed this; his listing of unique human capacities did not stop at language—it began there.[9] He was almost more interested in what language enabled than in it itself. For him, though language certainly allowed more complex communication between individuals, it also allowed us to become aware, through inner speech, of our thoughts. Through hearing ourselves think, through what he calls phonological (sound and language) imagery, we become aware of being aware. This meta-cognition, *the* human attribute, he suggested boldly, depends on speech and language, the unique and essential beginning before shared attention and awareness of others' mental states and our rich social interactions might become possible.

However, the contingencies between language and thought will not go away. Max Muller was quite clear: "No reason without speech. No speech without reason."[10]

There are, of course, many areas of profound thought independent of language; mathematics, for example, is expressed through other notations. Einstein, famously, related time and velocity through a visual thought experiment:

> Having arrived at results perfectly clear and satisfactory . . . when I try to express them in language, I feel I must begin by putting myself upon quite different another intellectual plane . . . the words of the language do not seem to play any role in my mechanism of thought. [My] elements of thought, [are] visual and muscular. Words have to be sought, laboriously, only in a secondary stage.[11]

Pinker mentions many others of similar mind, including Samuel Taylor Coleridge and Michael Faraday.[12] Nonlinguistic forms are not just carriers of ideas or thoughts either; fine art and music, especially classical music, provide unique depths of intellectual and emotional experience that resist linguistic analysis (though this does not prevent many from writing penetratingly and eloquently[13]). Dance, too, often reveals a form of thought and feeling in action, even though it may evade linguistic, conceptual analysis. I once asked Charlie Morrissey, who works with the British choreographer Siobhan Davies, about a piece he had developed.[14] What was it about? How did he interpret it? What was its meaning? Very kindly, he replied that he no articulatable thought nor any desire to analyze it in any linguistic, conscious form. Instead, the piece existed within some level of thought/ feeling and in action alone. Siobhan Davies describes her choreography as emerging from a sort of inner "soup" of feelings and explorations. She tries to harness these through dance, though she does not think, or imagine, or visualize movement; rather, she tries to express feeling first in and through movement almost without cognition and intellectual analysis. The magic is that when seeing her group develop and then perform their pieces, one does see feeling and movement joined in quiet statements of beauty and grace. But whether I can possibly see and feel what others do when viewing the same piece remains impenetrable; without language, the conceptual, shared semantic component may be absent.

Pinker gives the term "mentalese" to those areas of thought that, though they may require language to be expressed and communicated, exist in a form before or beyond language during their genesis and analysis. He gives many examples of thought occurring in an inner mentalese without language, even though it needs to be "worded" to be expressed, from an aphasic to a congenitally deaf child with a rich conceptual world without language. He details the limitations of language as an internal carrier of thoughts too. Taking English as his example, he shows how that language can be ambiguous.[15] How can a person have one thought when puns and double meanings in language are so frequent and often loved,[16] and when language itself can be logically imprecise?

> The representations underlying thinking, on the one hand, and the sentences in a language, on the other, are in many ways at cross-purposes. Any particular thought in our head embraces a vast amount of information. But when it comes to communicating to someone else, attention spans are show and mouths are slow. . . . A speaker can encode only a fraction of the message into words and must count on the listener to fill in the rest. . . .
>
> The language of thought . . . presumably . . . has symbols for concepts, and arrangements of symbols that to correspond to who did what to whom. . . . But compared with any given language, mentalese must be richer in some ways and simpler in others. It must be richer . . . in that several concept symbols must correspond to a given . . . word. . . . [and simpler in that some words] (like *a* and *the*) are absent, and information about pronouncing words . . . is unnecessary.[17]

Pinker dismisses any idea that thoughts might in some way be limited by their linguistic expression, so-called linguistic relativity, or the Sapir-Whorf hypothesis.[18] A question is how closely aligned Pinker's mentalese thoughts are to their means of expression or perception. Are mentalese thoughts about, say, form and content in fine art involving visual areas of the brain, and motor ones related to brush strokes on a canvas? Do thoughts to be expressed in words involve auditory and vocalization areas? Since the time Pinker wrote, thirty years ago now, there has been a huge development in the research on inner voices.[19] Recent work has explored its types and possible functions and its preservation or not in various types of aphasia. There is thus a vast literature on inner speech and its forms in people both with and without various impairments including and beyond aphasia, for instance, in autism, and in those with psychological problems leading to auditory verbal hallucinations.

One additional condition of interest in this regard may surprise. In *Seeing Voices*, Sacks mentions how on a visit to Martha's Vineyard where there is a high incidence of deafness, he saw an older woman on her porch. Though asleep, she was gesticulating; he speculated she was dreaming in sign language. He also wondered about the differing contents of speech and sign and whether sign, being more embodied, might be more emotionally expressive.[20]

Congenital deafness (often written "Deaf" to distinguish from those who lose hearing in later life, the "deaf") may appear an odd condition for insights into speech, whether vocalized or inner. However, people who live without sound and voice may have replaced the phonological loop thought so important by Passingham with internal visual signs. How, for instance, does a Deaf, signing actor learn their lines? Do they initially read text in words, then translate into sign, and then internalize those signs into a complex sequence? Though many children with hearing impairments now receive cochlear implants and so receive language through the spoken word, others still sign, sometimes in parallel with speech, so-called sign-supported speech (SSS) or simultaneous communication (SC).[21]

We now turn to the second form of nonsocial speech Vygotsky identified and the stage in children he thought preceded inner speech, so-called private speech. This is speaking to oneself in a self-dialogue that is not directed at anyone else but which they may hear. He thought children developed this for a short time, though others have suggested it may continue into adulthood. It has been considered to have several uses, including planning, organizing, and consolidating learning.[22]

Following Sacks's observation on Martha's Vineyard, Zimmermann and Brugger's main aim was to see whether people who were deaf used sign language in a way visible to others, which disclosed their use of an equivalent to private speech. In a sample of twenty-eight hearing and twenty-eight deaf adults (of whom twenty were congenitally deaf), they used a questionnaire to ask about their use of inner and private speech and its affective content.[23] Though both were used regularly, they found that deaf signers used both inner and private speech more, perhaps to compensate for any feelings of isolation. Signers were also aware that their signed soliloquy could be seen by others and consciously used it less in public spaces.

Alderson-Day and Fernyhough, who have studied inner speech and inner voices (particularly auditory hallucinations) for many years, quote a study of a semi-structured questionnaire assessing subjective experience during functional MRI by Delamillieure et al. who reported that only 17 percent of resting-state experiences described by their participants were language-based.[24]

Remarkably, two studies by Geva and colleagues have reported on inner speech and language skills in aphasia.[25] Most people with aphasia were impaired on both inner and overt speech, and performance in the two was closely correlated, but in a few, while inner speech was intact, overt speech was not, or vice versa, suggesting a partially discrete representation in the brain for the two domains.

This is also suggested by a unique report by Vercueil and Perrone-Bertolotti of a woman who experienced jargon-like inner speech during her epileptic seizures.[26] The experience of jargon in overt aphasia is well-documented, but few accounts exist of jargon in inner speech, not least because of the difficulties in comprehension and reporting. This subject did report her experience of jargon-like inner speech during seizures:

> Her written report mentioned the fact that during her seizures, even inner speech became incomprehensible, with the perception of an inner jargon which remained self-sustained throughout the seizure even though it sounded strange (she literally wrote: "Incomprehension of inner language (*thought is unintelligible*), and if I try to repeat inner language out loud, incomprehensible words come out (at any rate I don't understand them!).)"[27]

Note that she felt that as her inner speech was garbled, so were her thoughts. The authors concluded that there may be shared mechanisms in overt and inner speech, in contrast to the findings of Geva et al. It is not clear, however, that the two studies are mutually inconsistent, not least because the seizure discharge was unlikely to respect cortical boundaries in function.

Langland-Hassan et al. tested people with aphasia on silent rhyming, with the idea that this allows one to probe whether inner speech follows vocalized speech and recognizes rhyme or not.[28] They found that some people with aphasia were impaired but that this did not correlate with other tests of spoken, overt language and rhyming. Participants who could name objects and judge rhyme during vocalization were still bad at silent rhyme. Langland-Hassan et al. concluded this would suggest either, as did Vygotsky, that inner speech is to a degree a separate internalized system or that those with aphasia lacked the working memory to judge that two inner speech words rhymed.

Trying to understand the inner thoughts and feelings of others through silent and inner speech, and then to try to understand them in those with aphasia, as Sheila Hale and these studies have, requires patience and careful testing, and might still be like seeing the far side of the moon. But they all show that, just as the origins of speech may be social, so we are impelled to try to understand others, particularly those with impairments, and some of us are impelled to look even deeper into their internal worlds.

Acknowledgments

In preparing this book I met many medical and nonmedical professionals and an even larger number of people with various difficulties with speech. Since the names of those in the latter group are mostly withheld to preserve their anonymity, these acknowledgments—unusually—are to more unnamed people than named.

My interest was first awakened by Kate Heathcote, an ear, nose, and throat specialist at University Hospitals Dorset. She invited me to help her to distinguish vocal cord paralysis due to nerve problems from those due to structural problems in the vocal fold cartilages. She has an increasingly international reputation for nerve grafting in the former group to try to improve function. It was meeting patients with such paralysis that led me to understand for the first time some of the problems those with difficulties in speech production have.

Following this, I decided to approach groups of people with various voice problems from the perspective of an interested observer rather than a clinician. I emailed various online support groups, introduced myself and my purpose, and asked to be allowed to post an invitation for people to approach me if they felt comfortable doing so. I am immensely grateful that so many people did so and then trusted me to tell their narratives. All these accounts were then shown to each person or their partner or interlocutor for their approval. Whereas my previous books have relied on face-to-face conversations,[1] many of these support groups are international, so my reach could now extend further, with many conversations online via Zoom. This was also necessitated for part of the time by COVID restrictions.

I began, naturally, with those with vocal cord paralysis and must thank the Facebook Vocal Cord Paralysis Support Group and its members: https://

www.facebook.com/groups/189179717767866. People from the United States and Europe were very generous with their time for both conversation and then reading and correcting my prose.

My next exploration was to the community that lives with abnormal activation of the cords during speech. I am very grateful to many from the UK Spasmodic Dysphonia, Laryngeal Dystonia Help Group for their patience in relating their experiences.

When approaching those with cleft lip and palate I was fortunate to be able to ask the advice of Gareth Davies, who is cofounder of the European Cleft Organisation and was previously chief executive of the Cleft Lip and Palate Association of the United Kingdom, whom I had met previously through two brainchildren of James Partridge, Changing Faces and Face Equality International. Gareth very kindly pointed me toward various sites and also introduced me to executives of SmileTrain, USA, and South Africa. My thanks for discussions with Adina Lescher, Helena Curtis, and Caitlin Mescano. Most thanks, though, in relation cleft are to those who confided in me their experiences in such detail and with such candor.

I have written about Möbius syndrome several times, and each has featured Matthew Joffe, a wonderful man with the condition who has made himself available to others with Möbius so much over the years. Once more I am grateful for his short but essential reflections. The other person who spoke to me at length and over many years remains anonymous but has also made huge contributions over decades to matters Möbius and in the wider disability area and beyond. I have been fortunate to count each a colleague and friend.

I met several people with Parkinson's through the Parkinsons UK who must remain anonymous, and I extend my thanks to them and to Sarah Awde of Get Loud Therapy for explaining her method.

When considering the experiences of cerebral palsy, my thoughts turned to Kristian Møller Motke Martiny. Kristian's PhD was on cerebral palsy, and he now runs Enact:Lab in Copenhagen, a company whose purpose is to enable change in society's acceptance of diversities of whatever form. He generously arranged for me to meet Martin Nossell and Jacob Merrild, and they in turn set up a day's workshop in Copenhagen for discussion among a small community with cerebral palsy about their problems associated with speech and particularly others' responses to these. They also arranged for Jeppe Forchhammer and his PA to attend. These days in Copenhagen were spellbinding, and I hope I have captured something of the differing experiences of those with cerebral palsy.

I had met Minae Inahara and Michey Gillan Peckitt at a workshop in Hull in the United Kingdom and had always kept in touch. Minae had discussed her speech problems associated with cerebral palsy before, but she was kind enough to go into these in more depth and to discuss them with me face-to-face during a visit to Japan I made. Thanks to both for their discussions and hospitality.

Researching and writing can be hard work, but it is offset by the opportunities it gives to meet and enjoy remarkable people. Over the last few years I have been so fortunate to be able to do this as I listened to the narratives of all of the above.

There is an excellent speech and language therapy department at the University Hospitals Dorset, and I was very grateful to Deborah Broadbent for her interest and discussions. She emailed me about people I could interview post-stroke, and they were all kind enough to talk about their experiences. She also suggested I spend time with Catriona Lee and Faye Wright, who support those with chronic speech impairment after stroke. Faye also agreed to tell her own story, with her own name, for which I am very grateful. Thanks also to Faye for various pub breakfasts and lunches over the years. I had marveled at the account by Sheila Hale of her husband's aphasia after his stroke and am very grateful to her for sharing her insights over lunches in Twickenham. Professor Alex Leff of University College London was kind enough to straighten me out about aphasia at one stage and I am grateful for his time.

Deborah Broadbent suggested I see James and his wife Denise, without letting me know of his problem. I am so grateful to both of them for allowing me to listen to their extraordinary experiences and to piece it all together.

Leafing through artistic representations of talking online I came across Joseph Mann's work. Whereas most artists' portrayals of conversations focus on its meaning or context, whether a philosopher's soirée or a lover's tryst, Mann, rather than taking conversation itself for granted, seemed to introduce a sense of uncertainty and jeopardy. I am very grateful to him for allowing me to reproduce one such image on the cover of the book.

I have been fortunate to find someone to encourage me in my writing and who has shepherded this to publication. Thanks to Matthew Browne,

Anthony Zannino, Deborah Cantor-Adams, and Anne-Marie Bono of the MIT Press, and to Stephanie Sakson for assisting with the editing.

Nearly thirty years ago at a clinical meeting a woman there was without speech or facial expression. Some neurologists thought she was demented, but I could see this was not the case; she just could not communicate because of these two impairments. I went to see her and she tapped her story out on a Lightwriter, a sort of early laptop. Though at the time I did not know how I might ever write about her, I asked if I might tell her tale if the opportunity arose for me in the future. She tapped out, "Tell, Please."[2] There is sometimes the thought that people may not want to tell of their experiences, but on the whole I have not found this to be the case. There is a desire to share and to be understood. The privilege is to be trusted to do this.

Last, while writing can be solitary, when combined with a busy medical day job and other professional work it requires more than one person's efforts. I am fortunate that my wife supports and encourages me to write, while she carries more than her share of all the other stuff. To Sue, as ever, my gratitude and love.

Notes

Epigraph

1. George Eliot (then Mary Ann Evans) to John Sibree, Jr., 1848, in *George Eliot Letters*, vol. 1, ed. J. W. Cross (New York: Harper and Brothers, 1885). Available from Beverley Park Rilett, ed., *George Eliot's Life*, at GeorgeEliotArchive.org. When Evans accompanied her father to live in Foleshill, Coventry, in 1841, John Sibree Sr. and his wife were among the first people they met. He was an independent minister in a Coventry church. Their son was four years younger than Evans and they wrote long, intellectually challenging letters to each other. He later gave Evans lessons in Greek, while she challenged his religious beliefs and may have influenced his decision not to follow his father into the ministry.

Introduction

1. T. Cox, *Now You're Talking* (London: Bodley Head, 2018), 43. With a life of eighty years, and assuming we talk the same amount each day of those, then this works out as 17,000 per day, which sounds a lot.

2. Cox, *Now You're Talking*, 11.

3. R. Passingham, *What Is Special about the Human Brain?* (Oxford: Oxford University Press, 2018), 8–31.

4. R. Dunbar, *Grooming, Gossip and the Evolution of Language* (London: Faber and Faber, 1996), 115ff.

5. L. S. Vygotsky, *Thought and Language*, ed. and trans. Eugenia Haufmann and Gertrude Vahar (Cambridge, MA: MIT Press; New York: Wiley and Sons, 1962), 34. Original Russian edition published in 1934.

6. L. Wittgenstein, *Tractatus Logico-Philosophicus.* trans D. F. Pears and B. F. McGuinness (London: Routledge, 1961 [1922]); L. Wittgenstein, *Philosophical Investigations*, ed. G. E. M. Anscombe and R. Rhees, trans. G. E. M. Anscombe (Oxford: Blackwell,

1953). I do realize I have oversimplified his views on private language. There is a consideration of inner speech in the postscript to the present book.

7. O. Sacks, *An Anthropologist on Mars* (London: Picador, 1995), xiv.

8. H. Carel, *The Phenomenology of Illness* (Oxford: Oxford University Press, 2016), i.

9. There were reports that children with Möbius had an increased incidence of autism and learning difficulties in the past, but this reflected their physical and facial problems and, tragically, medical staff's inability to see beyond them.

10. Matthew Joffe, personal communication

11. Matthew Joffe, personal communication.

12. C. Bell, *The Hand: Its Mechanism and Vital Endowments as Evincing Design* (London: Pickering, 1833; reprint Brentwood: Pilgrims Press, 1979), 208–215. He also discussed the ingenuity of those born without hands in using their feet and even detailed a man born without arms who, despite this, was a murderer. His account goes on to ask why an orangutan cannot speak.

13. V. Kumar, P. L. Croxson, and K. Simonyan, "Structural Organisation of the Laryngeal Motor Cortical Network and Its Implication for Evolution of Speech Production," *Journal of Neuroscience* 36, no. 15 (2016): 4170–4181.

14. Passingham, *What Is Special about the Human Brain?*, 108.

15. C. Nolan, *Under the Eye of the Clock* (London: Weidenfeld and Nicholson, 1999); C. Brown, *My Left Foot* (London: Secker and Warburg, 1954).

Christopher Nolan initially found any controlled movement impossible for him:

> the first attempt to smile (when being welcomed at a new school) sent Joseph's [the name Nolan gave himself] body into spasm making facial muscles contort and arms and legs move violently as a clock work doll on being released to unwind. Joseph wished to thank his teachers, but spasm ruled out gratitude and instead grimaces. (16)

In the end, he learned to type, one letter at a time, with a manipulandum strapped to his head and the writhing and chaotic movements calmed by the drug Lioresal. Christy Brown famously found another way of expressing himself:

> Seeing [my] sister write, I wanted desperately to do what my sister was. Without thinking of knowing exactly what I was doing, I reached out and took the stick of chalk, with my left foot. I do not know why I used my left foot . . . that day it, apparently of its own volition, reached out. I held it tightly between the toes and acting on impulse, made a wild scribble. . . .
> I stiffened my body and tried once more. I shook, I sweated and strained every muscle. My hands were so tightly clenched that my fingernails bit into the flesh. I set my teeth so hard that I nearly pierced my bottom lip. I drew the letter "A."
> It had started. My mind had a chance of expressing itself. (15)

His childhood was relatively joyful, but as he grew older, he became more and more aware of his physical limitations and their effects. At age seventeen,

I felt myself coming to a greater, more persistent awareness of my own needs, and that in itself was pain enough. The pain went deeper when I also realized the impossibility of finding any adequate expression of such needs . . . no matter how I might overcome my physical limitations, my inner emotion life would never, could never, really be normal. It would have to lie bottled up inside me, suppressed instead of expressed.

My affliction was not after all incurable, but something else was, my lack of really normal human expression and relations. (78)

Chapter 1

1. These comprise muscles and ligaments and are more properly called "folds," but "cords" will be used since this term is in common usage.

2. Discussing the evolution of the human hand, in 1833, Bell lined up drawings of the hands of a large series of primates and saw little difference. Our greater use of the hand for manipulation and action depended, he reasoned, on development not of the hand but of the brain to control it in new ways. Charles Bell, *The Hand: Its Mechanism and Vital Endowments as Evincing Design* (London: Pickering, 1833; reprint Brentwood: Pilgrims Press, 1979), 208–215.

3. For videos of the vocal cords moving from above, see "Vocal Folds—Pitch Physics Demo," YouTube, https://www.youtube.com/watch?v=wYnPA7IXFIU; "Speech-Language Pathology: The Vocal Cords in Action," YouTube, https://www.youtube.com /watch?v=y2okeYVclQo. For a short animation of the action of the muscles involved, see "Vocal Cord Muscles (Adduction and Abduction)," YouTube, https://www.youtube .com/shorts/Ohg3RIDXTrM. The larynx can also move up and down in the neck.

4. When running with an athletic club we prided ourselves on being able to speak in normal sentences during our long, slower training runs.

5. "Zzz" refers to the use of "z" in a word and phoneme. UK English pronunciation of the isolated letter "z" is "zed," so less voiced than the US "zee."

6. S. Ahmad, A. Muzamil, and M. Lateel, "A Study of Incidence and Aetiology of Vocal Cord Paralysis," *Indian Journal of Otolaryngology and Head Neck Surgery* 54, no. 4 (2002): 294–296, https://doi.org/10.1007/BF02993746.

7. C. Y. Lo, K. F. Kwok, and P. W. Yuen, "A Prospective Evaluation of Recurrent Laryngeal Nerve Paralysis during Thyroidectomy," *Archives of Surgery* 135, no. 2 (2000): 204–207, https://doi.org/10.1001/archsurg.135.2.204; G.-R. Joliat, V. Guarnero, N. Demartines, V. Schweizer, and M. Matter, "Recurrent Laryngeal Nerve Injury after Thyroid and Parathyroid Surgery: Incidence and Postoperative Evolution Assessment," *Medicine* 94, no. 17 (2017): e6674, https://doi.org/10.1097/MD.0000000000 006674; V. K. Dhillon, E. Rettig, S. I. Noureldine, D. J. Genther, A. Hassoon, M. G. Al Kahdem, O. B. Ozgursoy, et al., "The Incidence of Vocal Fold Motion Impairment after Primary Thyroid and Parathyroid Surgery for a Single High-Volume Academic Surgeon Determined by Pre- and Immediate Post-Operative Fiberoptic Laryngoscopy," *International Journal of Surgery* 56 (2018): 73–78.

8. Surgeons usually do pronounce their work a success.

9. A neurological colleague told me of his duodenal ulcer bleed. He realized something was wrong when he started passing thick, black stool from the blood loss, known as melena. He had two days of busy clinics and felt okay, so carried on. Once home and a keen athlete, he then went for a run to test himself. When he was endoscoped and told that the ulcer had eroded a blood vessel, they also informed him he had a 20 percent mortality rate. He did not tell them of the run. It was so deeply ingrained in him, like Armstrong, that if he could run he felt he was healthy. Despite two endoscopy sessions and several sleepless days and nights in hospital, he said the worst thing was the well-deserved telling off from his wife for that run.

10. I interviewed Elaine's therapist at the suggestion of and with the permission of Elaine.

11. J. Cole, *About Face* (Cambridge, MA: MIT Press, 1998), 25.

Chapter 2

1. To pass oneself off as being no different from one's peers in the face of disfigurement or impairment is often the aim. See S. Gilman, *Creating Beauty to Cure the Soul* (Durham, NC: Duke University Press, 1998).

2. Her spoken and written words have very different sentence lengths and complexity, reflected here.

Chapter 3

1. In Ondine's curse individuals *do* stop breathing when asleep. Congenital central hypoventilation syndrome is rare and noted in the neonatal unit after delivery. It is named after a mythical tale in which a heartbroken water nymph curses her unfaithful husband so that he would stop breathing should he ever fall asleep.

2. Politicians, notoriously, tailor not only their words but their speech patterns and accents to their audience.

3. In the movie *The Little Mermaid*, she is depicted as being willing to do anything to be with Prince Eric, even giving up her voice to become human. Ron Clements and John Musker, directors, *The Little Mermaid* (Walt Disney Pictures, 1989).

Chapter 4

1. When the lining of the pharynx and larynx is anesthetized, people often think they are not breathing in and out properly, because they suddenly become aware

of not feeling airflow through the throat. This sensory feedback guides breathing, though we are not normally aware of it.

2. B. D. Berman, M. Hallett, P. Herscovitch, and K. Simonyan, "Striatal Dopaminergic Dysfunction at Rest and during Task Performance in Writer's Cramp," *Brain* 136 (2013): 3645–3658, https://doi.org/10.1093/brain/awt282.

3. J. Konsak and G. Abbruzzese, "Focal Dystonia in Musicians: Linking Motor Symptoms to Somatosensory Dysfunction," Frontiers in Human Neuroscie*nce* 7 (2013): 297, https://doi.org/10.3389/fnhum.2013.00297; E. Altenmüller, "Focal Dystonia: Advances in Brain Imaging and Understanding of Fine Motor Control in Musicians," *Hand Clinics* 19 (2003): 523–538, https://doi.org/10.1016/S0749-0712(03)00043-X.

4. E. Altenmüller and H.-C. Jabusch, "Focal Dystonia in Musicians: Phenomenology, Pathophysiology and Triggering Factors," *European Journal of Neurology* 17 (July 2010): 31–36, https://doi.org/10.1111/j.1468-1331.2010.03048.x; E. Altenmüller, V. Baur, A. Hofmann, V. K. Lim, and H.-C. Jabusch, "Musician's Cramp as Manifestation of Maladaptive Brain Plasticity: Arguments from Instrumental Differences," *Annals of the New York Academy of Sciences* 1252 (2012): 259–265, https://doi.org /10.1111/j.1749-6632.2012.06456.x; A. Chitkara, T. Meyer, A. Keidar, and A. Blitzer, "Singer's Dystonia: First Report of a Variant of Spasmodic Dysphonia," *Annals of Otology, Rhinology, and Laryngology* 115, no. 2 (February 2006): 89–92, https://doi.org /10.1177/000348940611500201.

5. L. A. Halstrad, D. M. McBroom, and H. S. Bonilha, "Task-Specific Singing Dystonia: Vocal Instability That Technique Cannot Fix," *Journal of Voice* 29, no. 1 (January 2015): 71–78, https://doi.org/10.1016/j.jvoice.2014.04.011.

Chapter 5

1. She thinks her reactions to Botox are a consequence of what had happened previously at the clinic with the two consultants.

2. Disability Discrimination Act 1995, https://www.legislation.gov.uk/ukpga/1995 /50/contents.

3. Kristina Simonyan, Dystonia and Speech Motor Control Laboratory, https:// simonyanlab.meei.harvard.edu.

4. L. C. O'Flynn, S. J. Frucht, and K. Simonyan, "Sodium Oxybate in Alcohol-Responsive Essential Tremor of Voice: An Open-Label Phase II Study," *Movement Disorders* 38, no. 10 (October 2023): 1936–1944, https://doi.org/10.1002/mds.29529. See also "Brain Computer Interfaces: Decision Making and Dystonia Treatment," Dystonia and Speech Motor Control Laboratory, https://simonyanlab.meei .harvard.edu/research/brain-computer-interfaces-decision-making-and-dystonia -treatment.

5. Though rare, there may be a disproportionate number of teachers and singers with the condition. Given their extensive use of the voice professionally, this might be expected, though whether it might be classified as an occupational disease is another matter. Robert F. Kennedy Jr. is one of the higher profile people with it at present.

Chapter 6

1. Great Ormond Street Hospital for Children, NHS Foundation Trust, "Cleft Lip and Palate," https://www.gosh.nhs.uk/conditions-and-treatments/conditions-we-treat /cleft-lip-and-palate; Mayo Clinic, "Cleft Lip and Cleft Palate, Symptoms and Causes," https://www.mayoclinic.org/diseases-conditions/cleft-palate/symptoms-causes /syc-20370985.

2. I did debate whether I should reproduce them. The anonymized participants wanted their stories told.

3. One of our daughters had talked all day when aged three or four. Suddenly, she stopped: "I've run out of words." I saw her next with my big Oxford dictionary on her lap. She could not read but somehow knew that was where the words were.

4. Though she worries a little about not hearing any alarms.

5. Changing Faces, "Disfigurement in the UK," May 2017, https://www.changing faces.org.uk/wp-content/uploads/2021/05/disfigurement-in-the-uk-report-2017.pdf.

6. Changing Faces, "My Visible Difference," https://www.changingfaces.org.uk/wp -content/uploads/2021/01/CHANGING-FACES-Report-My-Visible-Difference.pdf.

7. Public Health England, "Disability and Domestic Abuse: Risk, Impacts, and Response," November 2015, https://assets.publishing.service.gov.uk/media/5a8066 73ed915d74e622e3c8/Disability_and_domestic_abuse_topic_overview_FINAL.pdf.

8. Ann Craft Trust, "Disability & Domestic Abuse," posted by Abra Millar in Safe-guarding Adults, November 6, 2018, https://www.anncrafttrust.org/disability-dome stic-abuse.

9. Changing Faces, "Your Legal Protection from Discrimination at Work as Someone with a Visible Difference," https://www.changingfaces.org.uk/advice-guidance/work ing-when-you-have-visible-difference/equality-act-protection-discrimination-work.

10. ADA.gov, "Introduction to the Americans with Disabilities Act," https://www .ada.gov/topics/intro-to-ada.

Chapter 7

1. In the past, some children with Möbius were thought autistic, but this was found to be due to their difficulties in development and in interpersonal relatedness

secondary to their facial disfigurement. See J. Cole with H. Spalding, *The Invisible Smile* (Oxford: Oxford University Press, 2008), 180–184.

2. Cole and Spalding, *The Invisible Smile*, considered the lives and experiences of those with Möbius.

3. Jo is nonbinary and uses the pronouns "they/them."

4. An example of this is one heavy metal track that during the fadeout is supposed to have a satanic message when played backwards (those were the days). When you first hear this, it sounds like the gobbledy-gook it is, but if you are primed to the words by seeing them written down, then you hear them clearly on the backward-played track. What we hear, like much of our sensory information, depends on top-down predictive control.

5. Jo met someone with Möbius who lived in Sweden. "He was one of the saddest people; completely wooden, with no body language, like a puppet. I met him in France; I hope he learned something from them. If I had not lived in Romania, I don't know how I would have turned out."

6. M. Merleau-Ponty, "Eye and Mind," in *The Primacy of Perception,* ed. James Edie, trans. Carleton Dallery (Evanston, IL: Northwestern University Press, 1964), 159–190.

Chapter 8

1. "Enjoyed talking?" she added on reading a draft.

2. James Parkinson's original observations on his eponymous condition were made on the streets of London watching people walk, rather than by formal examination.

3. "Once I did see that I really can't look cross when I need to look cross—I was driving and was cut up badly by a chap in a posh car and I did my road rage face at him. Then I whipped the sun visor down and looked at my mild, equable, sweet-tempered face in the makeup mirror."

4. "Parkinson's appears to develop without the person knowing until about 80 percent of the relevant chemical in the brain has gone. My whole time as head of department was probably colored by it, unknown to begin with, then known. It may explain my imposter syndrome."

5. Ever rigorous, she corrected her use of various drugs. "Strictly, my consultant did the trying."

6. Ever questioning her own narrative, she wrote about this part and her use of the term "slipping away." "Is there a point to this? Oh yes, PD is degenerative, so I, um, degenerated."

7. Lee Silverman Voice Treatment (LSVT) is a form of intensive voice therapy for those with Parkinson's disease and other neurological conditions that affect the voice.

8. It was not that she was softer or less emphatic; with this came changes in the range of her expression. "Being jokey is difficult, and then one is viewed as serious, or down with the condition, or stand-offish." The link between outer expression and inner experience is no longer a given. "At least I can be flippant and jokey in an email—telling jokes is really tricky with voice problems."

Chapter 9

1. CerebralPalsy.org, "Prevalence of Cerebral Palsy," https://www.cerebralpalsy.org /about-cerebral-palsy/prevalence-and-incidence.

2. Enact:Lab works with and for people to create the agency to change and then solve the structural problems of society. Through scientific research, open science, collaboration, knowledge sharing, and interdisciplinary processes, they aim for the creation of agency and social change that can lead toward sustainable solutions on problems in society. Their work originated in disability, but they have broadened to work with companies and public bodies. "To create real change . . . a new approach is needed that is deeply rooted in building agency, centering on lived experience, communities, and cultivating inclusive social dynamics" (https://enactlab.com).

3. Enact:Lab's diversity specialist and researcher Emilie Lund Palsøe was the main coordinator of this project and helped set up the process.

4. Their study also plans to interview people without any speech impairment at a later date.

5. The interviews were also recorded for further analysis.

6. I was once at a very well-to-do party in an affluent country house in the south of England. A girl came up to me and, to start conversation, asked where I was from. "Birmingham," I replied. She turned and without a backward glance walked off. I knew Birmingham was not fashionable, but still.

7. Jacob is quite famous in the media in relation to disability. He complained that when he and Kristian (also a media aficionado) are on together, Kristian can talk too much and too fast. Sometimes to get his own back in front of an audience, he will use Kristian to translate his sentences, and then when Kristian says the words, Jacob will deny ever saying it like that, just to set Kristian up.

Chapter 10

1. This is a quote from Martin's master's dissertation, "An autoethnographic exploration of people with communication disabilities."

2. Many adolescent males are embarrassed about their voices. A friend said his fifteen-year-old son discovered the power of speech only when he had his first girlfriend.

3. To complicate matters, the Danish and English alphabet differ, with the Danish adding Æ, Ø, and Å and not using Q, X, or Z. In the transcriptions, English language spellings are used.

Chapter 11

1. At the Johnson Space Center, in Houston, Texas, we were being shown round when we got to a big pool. Sunk at the bottom was a shuttle mock-up. The astronauts, in full suits, were submerged, practicing their tasks ahead of launch. I asked one of their teachers how they helped the guys learn. "We show them what they need to do, let them see how they want to do it, and then help them do it better their way." Good medical rehabilitation in a nutshell, I thought.

2. M. Merleau-Ponty, *Phenomenology of Perception*, trans. Colin Smith (New York: Humanities Press; London: Routledge & Kegan Paul, 1962), 174–199.

Chapter 12

1. Strictly, to have reduced speech should be termed "dysphasia," with "aphasia" reserved for those without speech entirely. However, this term can be confused with "dysphagia," problems with swallowing, so "aphasia" is used widely, at least in the United Kingdom.

2. Centers for Disease Control and Prevention, "Stroke Facts," https://www.cdc.gov /stroke/facts.htm. UK figures are 100,000 strokes per year with 1.3 million survivors. Stroke Association, "Stroke Statistics," https://www.stroke.org.uk/stroke/statistics.

3. C. J. Price, "A Review and Synthesis of the First 20 Years of PET and fMRI Studies of Heard Speech, Spoken Language and Reading," *Neuroimage* 62, no. 2 (2012): 816–847.

4. Stroke Awareness Foundation, "Stroke Facts & Statistics," https://www.stroke.org .uk/stroke/statistics; National Aphasia Association, "Aphasia Statistics, https://www .aphasia.org/aphasia-resources/aphasia-statistics. Around 350,000 people live with aphasia in the United Kingdom.

5. Before diagnosing aphasia, it is necessary to exclude problems with hearing, and with movement and control of the vocal apparatus (an apraxia). Only

then can a diagnosis of a linguistic communication impairment be diagnosed, caused by damage in the areas of the brain that control language expression and comprehension.

6. D. Howard and F. M. Hatfield, *Aphasia Therapy: Historical and Contemporary Issues* (Hove: Erlbaum, 1987), quote from S. Hale, *The Man How Lost His Language* (London: Jessica Kingsley, 2007), 93.

7. See ASHA.org for classification of aphasia, at asha.org/asha/#q=classification%20 of%20aphasia&sort=relevancy, which develops the classification of H. Goodglass and E. Kaplan, *The Assessment of Aphasia and Related Disorders* (Philadelphia: Lea and Febiger, 1972).

8. Pierre Paul Broca and Carl Wernicke are renowned figures in the history of neurology. Broca described, in 1861, two cases of people with nonfluent aphasia, who somewhat conveniently had died and had been submitted to autopsy. Broca found the areas of brain involved were the same left second and third frontal convolutions, lateral to the motor cortex. Previously, language production had been localized to the frontal lobes, but not so precisely. It was an important example of how different parts of the cerebral cortex had different functions. See J. M. S. Pearce, "Broca's Aphasiacs," *European Neurology* 61 (2009): 183–189. https://doi.org/10.1159/000189272.

Wernicke's work was arguably a little more nuanced. He did describe an area subsequently named for him, where language comprehension was located, in the left posterior temporal lobe, but he was also aware of the importance of coordinated actions between areas, and so subcortical connections between cortical areas would prove crucial. As Eggert suggests, over a period of thirty years or so, Wernicke looked for connections between sensory and motor speech areas, and thought that areas of frontal, temporal, and parietal cortices were linked functionally. See G. T. Eggert, *Wernicke's Works on Aphasia: A Sourcebook and Review* (The Hague: Mouton, 1977). Wernicke also introduced the idea that speech had to be self-monitored. As we speak, we have to monitor the words used to make sure they match our stored and accepted words and their sounds.

9. W. J. M. Levelt, "Models of Word Production," *Trends in Cognitive Science* 3, no. 6 (1999): 223–232.

10. This is due to damage to the left occipital or posterior lobe in the brain, leading to an inability to see to the right side of visual space with either eye, a right homonymous hemianopia.

11. Cotard's syndrome.

12. It is likely that he has visual form agnosia, leading to deficits in recognition of objects when seeing them, which is a function related to the lateral occipital cortex, which may in turn be associated with the posterior occipital cortex damage leading to the hemianopia.

13. Since Steve has lost visual processing in the left occipital or posterior part of the brain, and speech, suggesting a left inferior frontal deficit, the brain areas between those two are likely to be affected also. Reading emerged culturally and involves several areas also involved in other functions, including spoken language and object recognition. It requires recognition of combinations of letters into their sounds and recognition of familiar words into their mental representations. These occur in a network of regions mainly located in the left hemisphere including the occipito-temporal, temporo-parietal, and inferior frontal cortices. Steve's stroke affected these first two at the least. See International Dyslexia Association, "Dyslexia and the Brain," https://dyslexiaida.org/dyslexia-and-the-brain; S. Dehaene and L. Cohen, "Cultural Recycling of Cortical Maps," *Neuron* 56, no. 2 (2007): 384–398, https://doi.org/10.1016/j.neuron.2007.10.004; C. J. Price, "A Review and Synthesis of the First 20 Years of PET and fMRI Studies of Heard Speech, Spoken Language and Reading," *Neuroimage* 62, no. 2 (2012): 816–847.

14. Bill preferred to use his real name. His aphasia seems mixed, with some fluent and nonfluent elements. Here as elsewhere, his new speech patterns have been preserved.

15. One person said that her "brain was like a washing machine, with all the words rushing around." She "could see them but not get hold of one to use."

16. G. Wint, *The Third Killer: Meditations after a Stroke* (London: Chatto and Windus, 1965), 16. He recovered after three to four years.

17. Wint, *The Third Killer*, 15.

18. C. Hörnsten, H. Lövheim, P. Nordström, and Y. Gustafson, "The Prevalence of Stroke and Depression and Factors Associated with Depression in Elderly People with and without Stroke," *BMC Geriatrics* 16, no. 1 (October 2016): 174, https://doi.org/10.1186/s12877-016-0347-6.

19. S. A. Ashaie, R. Hurwitz, and L. R. Cherney, "Depression and Subthreshold Depression in Stroke-Related Aphasia," *Archives of Physical Medicine and Rehabilitation* 100, no. 7 (2019): 1294–1299.

20. He was one of the few people interviewed who had found anything positive about their condition.

21. S. T. Pendlebury and P. M. Rothwell, "Prevalence, Incidence, and Factors Associated with Pre-Stroke and Post-Stroke Dementia: A Systematic Review and Meta-Analysis," *Lancet Neurology* 8 (2009): 1006–1018; see also commentary by M. Dichgans, "Dementia Risk after TIA and Stroke," *Lancet Neurology* 18, no. 3 (2019): 223–225, https://doi.org/10.1016/S1474-4422(18)30497-6.

22. G. M. Savva and C. M. Blossom, "Stephan and the Alzheimer's Society Vascular Dementia Systematic Review Group: Epidemiological Studies of the Effect of Stroke on Incident Dementia—A Systematic Review," *Stroke* 41, no. 1 (2010): e41–e46, https://doi.org/10.1161/STROKEAHA.109.559880.

23. S. T. Pendlebury, "Dementia in Patients Hospitalized with Stroke: Rates, Time Course, and Clinico-Pathologic Factors," *International Journal of Stroke* 7, no. 7 (2012): 570–581. https://doi.org/10.1111/j.1747-4949.2012.00837.x. This figure was dwarfed by that quoted on a website, which suggested that almost a quarter of people who have had a stroke will go on to develop dementia after about three to six months, which seems high, but of course case mix will have differed between groups. Alzheimer's Society, "Understanding Why Dementia Can Occur after Stroke," https://www.alzheimers.org.uk/research/our-research/research-projects/under standing-why-dementia-can-occur-after-stroke.

24. D. W. Desmond, J. T. Moroney, M. C. Paik, M. Sano, J. P. Mohr, S. Aboumatar, C. L. Tseng, et al., "Frequency and Clinical Determinants of Dementia after Ischemic Stroke," *Neurology* 54 (2000): 1124–1131.

25. H. L. Lin, C.-F. Tsai, S.-P. Liu, C.-H. Muo, and P.-C. Chen, "Association between Aphasia and Risk of Dementia after Stroke," *Journal of Stroke and Cerebrovascular Diseases* 31, no. 12 (2022): 106838, https://doi.org/10.1016/j.jstrokecerebrovasdis .2022.106838.

Chapter 13

1. In those days, GPs commonly had evening clinics.

2. A. Kopit, Preface to the play *Wings* (1978). Quoted from S. Hale, *The Man Who Lost His Language* (London: Allen Lane, 2002), 31. Quotation marks added by the present author. Arthur Kopit conceived the play after his father had a stroke. The play was very well received, won many awards, and has been revived several times, including once as a musical.

3. L. McWhirter, N. Miller, C. Campbell, I. Hoeritzauer, A. Lawton, A. Carson, and J. Stone, "Understanding Foreign Accent Syndrome," *Journal of Neurology, Neurosurgery and Psychiatry* 90, no. 11 (2018), http://dx.doi.org/10.1136/jnnp-2018-319842.

4. Alex Leff, a professor of cognitive neurology, neurotherapeutics group leader at University College London Queen Square Institute of Neurology, and an expert in the rehabilitation of those with speech impairments, told me that he no longer asks people with aphasia about the preservation of thought, in part because he questions how they might be aware of such deficits.

Chapter 14

1. Real names are used here by preference.

2. S. Hale, *The Man Who Lost His Language* (London: Allen Lane and Kingsley, 2007 [2002]), 197; quote from A. R. Damasio, *The Feeling of What Happens, Body, Emotion and the Making of Consciousness* (San Diego: Harcourt, 1999).

3. J. Fonseca, J. J. Ferreira, and I. Pavoe Martins, "Cognitive Performance in Aphasia Due to Stroke: A Systematic Review," *International Journal on Disability and Human Development* 16, no. 8 (2017): 127–139.

4. H. Hanane El Hachioui, E. G. Visch-Brink, H. F. Lingsma, M. W. M. E. Van De Sandt-Koenderman, D. W. J. Dippel, P. J. Koudstaal, and H. A. M. Middelkoop, et al., "Nonlinguistic Cognitive Impairment in Poststroke Aphasia: A Prospective Study," *Neurorehabilitation and Neural Repair* 28, no. 3 (2014): 273–281, https://doi .org/10.1177/1545968313508467.

5. C. V. Marinelli, S. Spaccavento, A. Craca, P. Marangolo, and P. Angelelli, "Different Cognitive Profiles of Patients with Severe Aphasia," *Behavioural Neurology* (2017), http://dx.doi.org/10.1155/2017/3875954.

6. H. Akkad, T. M. H. Hope, C. Howland, S. Ondobaka, K. Pappa, D. Nardo, J. Duncan, et al., "Mapping Language and Non-Language Cognitive Deficits in Post-Stroke Anomic Aphasia," working paper, http://dx.doi.org/10.1101/2021.02.15.431293.

7. Around nine months after I saw them, Rita phoned to say that Ron had died peacefully.

Chapter 15

1. Names here have been anonymized.

2. I reminded her of Chekhov's reflection, "Dogs are wonderful people."

3. S. Hale, *The Man Who Lost His Language: A Case of Aphasia*, 2nd ed. (London: Jessica Kingsley, 2007).

4. She has written a biography of Titian: S. Hale, *Titian: A Life* (London: HarperPress, 2012).

5. Some of the finest long and rich accounts of neurological conditions are from Luria, who studied his subjects for thirty years. See A. R. Luria, *The Man with a Shattered World* (Cambridge, MA: Harvard University Press, 1987), and *The Mind of a Mnemonist: A Little Book about a Vast Memory* (Cambridge, MA: Harvard University Press, 1986). I am not so modest as to fail to mention that I have published two biographies of the same man with a very rare loss of touch and movement sense, one twenty-five years after the other and which covered forty years of his experience: *Pride and a Daily Marathon* (London: Duckworth; Cambridge, MA: MIT Press, 1995 [1991]) and *Losing Touch* (Oxford: Oxford University Press, 2016). We are still researching together, nearly forty years after we met.

6. *The Blue Lagoon* (1949).

7. Sheila told me that John's Italian speech was not very good. When in Italy he used the wrong tenses and made other mistakes, but that they loved him for his

charm, eloquence, and Englishness, nevertheless. "He was never very good at foreign language. He could read all foreign languages but was less interested in speaking them well." Not speaking the language well was obviously no bar to being a professor of Italian.

8. Hale, *The Man Who Lost His Language*, 102.

9. Hale, *The Man Who Lost His Language*, 109.

10. Hale, *The Man Who Lost His Language*, 186.

11. Hale, *The Man Who Lost His Language*, 195, 196.

12. Hale, *The Man Who Lost His Language*, 202.

13. Hale, personal communication, February 13, 2023.

14. Hale, *The Man Who Lost His Language*, 169.

15. Hale, *The Man Who Lost His Language*, 202.

16. Hale, *The Man Who Lost His Language*, 199.

17. Hale, *The Man Who Lost His Language*, 200.

18. Hale, *The Man Who Lost His Language*, 196–197.

19. Hale, *The Man Who Lost His Language*, 200.

20. Hale, personal communication, February 13, 2023.

21. I wondered, What if an actor portrayed John's speech patterns, or he was replayed a passage from one of his own documentaries? Would he have recognized it as himself?

22. Hale, personal communication, February 13, 2023.

23. Hale, *The Man Who Lost His Language*, 203. It might have been interesting also to play back to him his speech from before the stroke.

24. Hale, *The Man Who Lost His Language*, 203.

25. Hale, *The Man Who Lost His Language*, 78.

26. Hale, *The Man Who Lost His Language*, 210.

27. Hale, *The Man Who Lost His Language*, 213.

28. Hale, *The Man Who Lost His Language*, 171.

29. This flip into easy fluency from difficult meaningful language has parallels with some movement tasks. If we rotate the arms in front of us, either both clockwise, or symmetrically, so the left, say, goes clockwise and the right counterclockwise, as we speed up, we degrade or adopt the easier symmetrical pattern unconsciously in both

conditions, choosing speed and ease over instruction. Dropping into fluency seems a similar, less effortful, and maybe unconscious change. Hale, *The Man Who Lost His Language*, 175.

30. Hale, *The Man Who Lost His Language*, 213.

31. He died, in his sleep, five years after the stroke, in 1997.

32. D. Howard and F. M. Hatfield, *Aphasia Therapy: Historical and Contemporary Issues* (London: Routledge and Taylor, 1987); Hale, *The Man Who Lost His Language*, 93.

33. G. Wint, *The Third Killer: Meditations after a Stroke* (London: Chatto and Windus, 1965), 30.

34. Wint, *The Third Killer*, 97.

35. An image that came to mind when working with those with fluent aphasia, perhaps surprisingly, was that of the artist Jeremy Deller's 2016 street art *We're Here Because We're Here*. Asked to commemorate 100 years since the Battle of the Somme in World War I, he arranged, completely secretly, for groups of amateur actors to appear in period army uniforms in public spaces (station concourses, university squares, and other locations) and to stand and mill around silently. If approached, they just handed out a card of a soldier with his name, regiment, and place where he was killed. Imagining the soldiers unaware of their fate transported to the same space as we now stood moved many onlookers to tears.

36. The work of speech and language therapists in assisting with language reacquisition and with countless other means of support cannot be underestimated. I am only too aware that my present exploration of narratives in those speech problems may seem naïve and simplistic to many in this field.

37. J. M. Lam and W. P. Wodchis, "The Relationship of 60 Disease Diagnoses and 15 Conditions to Preference-Based Health-Related Quality of Life in Ontario Hospital-Based Long-Term Care Residents," *Medical Care* 48 (2010): 380–387, https://doi.org/10.1097/MLR.0b013e3181ca2647.

Chapter 16

1. L. Vygotsky, *Thought and Language*, ed. A. Kozulin (Cambridge, MA: MIT Press, 1986).

2. "We are aware of ourselves, for we are aware of others, and in the same way as we know others; and this is as it is because in relation to ourselves, we are in the same [position] as others are to us." See L. Vygotsky, "Consciousness as a Problem of Psychology of Behaviour," *Soviet Psychology* 17 (1979): 29–30. Quoted from introduction by A. Kozulin, in Vygotsky's *Thought and Language*, xxiv.

3. Though his great work was called *Thought and Language*, his own preface began that the study was of one the most complex problems in psychology, the interrelation of thought, language, and speech.

4. Vygotsky, *Thought and Language*, 34–35.

5. Vygotsky, *Thought and Language*, 82–84.

6. Vygotsky, *Thought and Language*, 252–253.

7. A somewhat similar presentation was published by Wagner and Begaz. A forty-seven-year-old man, brought to the hospital by his wife, was without any complaint himself, and appeared "well dressed, talkative, and motivated." His wife was disturbed because he was talking less, dressing sloppily, sleeping longer, and not going to work. The doctors found little on detailed examination except that he was slightly slow to respond and had a flattened mood. They also noted that he did not speak unless spoken to, though his responses were fluent and appropriate. A scan showed a stroke in the left basal ganglion. He recovered over the next two months. One of Wagner and Begaz's conclusions was that "spouses have a unique insight into their partner's behavior and their concerns must be taken seriously." The case also highlights that neurologists home in on deficits in sensation, movement, speech, and action, but less on the aspects of emotional experience and social relatedness so disturbed here. See S. Wagner and T. Begaz, "Basal Ganglion Stroke Presenting as Subtle Behavioural Change," *BMJ Case Reports* 2009 (May 2009), https://doi.org /10.1136/bcr.10.2008.1139.

8. She added that until they started to talk in more depth, no one seemed to show any real grasp of James's situation (or therefore of hers). The present process had helped them to understand and to explain it to their friends and colleagues more easily. When they first received the diagnosis, they had little support.

9. P. Kailash, C. Bhatia, and C. D. Marsden, "The Behavioural and Motor Consequences of Focal Lesions of the Basal Ganglia in Man," *Brain* 117, no. 4 (August 1994): 859–876, https://doi.org/10.1093/brain/117.4.859.

10. R. S. Marin, "Apathy: Concept, Syndrome, Neural Mechanisms, and Treatment," *Seminars in Clinical Neuropsychiatry* 1, no. 4 (October 1996): 304–314, https://doi .org/10.1053/scnp00100304 PMID: 10320433. Emotional distress was included because some people show such a problem after this and after torture and solitary confinement.

11. C. Burgon, S. Goldberg, V. van der Wardt, and R. H. Harwood, "Experiences and Understanding of Apathy in People with Neurocognitive Disorders and Their Carers: A Qualitative Interview Study," *Age and Ageing* 52, no. 3 (March 2023): afad031, https://doi.org/10.1093/ageing/afad031.

12. R. Levy and V. Czernecki, "Apathy and Basal Ganglia." *Journal of Neurology* 253 (December 2006): 54–61, https://doi.org/10.1007/s00415-006-7012-5.

13. Another time he also said that he was aware of his situation and had an inner voice or monitor. He could be in the moment and be aware of being in the moment at the same time, so to speak.

14. Levy and Czernecki, "Apathy and Basal Ganglia."

15. J. Lennon and P. McCartney, "Julia." John's mother died in a car accident when John was eighteen. Lennon is supposed to have "borrowed" the first half of that sentence from Kahlil Gibran's *Sand and Foam*, which in the original verse reads, "Half of what I say is meaningless, but I say it so that the other half may reach you." Read more: Lorenzo Tanos, "The Tragic Death of John Lennon's Mother," Grunge, March 29, 2021, https://www.grunge.com/368139/the-tragic-death-of-john-lennons -mother. It appears to be the only Beatles song Lennon wrote and performed on his own. He wrote lots afterwards of course.

16. R. Murphy, *The Body Silent* (New York: Henry Holt, 1987), 78.

Chapter 17

1. J. Cole, *About Face* (Cambridge, MA: MIT Press, 1998), 33.

2. C. Gilligan, *In a Different Voice: Psychological Theory and Women's Development* (Cambridge, MA: Harvard University Press, 2003 [1982]).

3. E. D. Ross and M. Monnot, "Neurology of Affective Prosody and Its Functional-Anatomic Organization in Right Hemisphere," *Brain and Language* 104, no. 1 (January 2008): 1–74, https://doi.org/10.1016/j.bandl.2007.04.007.

4. O. Sacks, "The President's Speech," in *The Man Who Mistook His Wife for a Hat* (London: Picador, 1985), 76–80.

5. J. Claypole, *Words Fail Us: In Defence of Disfluency* (London: Profile Books/Well-come, 2021), 181.

6. N. Gerrard, *What Dementia Teaches Us about Love* (London: Allen Lane, 2019).

7. Gilligan, *In a Different Voice,* quoted in Gerrard, *What Dementia Teaches Us about Love.*

8. Gerrard, *What Dementia Teaches Us about Love,* 25.

9. Alzheimer's Society, "How to Communicate with a Person with Dementia," https://www.alzheimers.org.uk/about-dementia/symptoms-and-diagnosis/symptoms /how-to-communicate-dementia.

10. M. Ray, T. Dening, and B. Crosbie, "Dementia and Hearing Loss: A Narrative Review," *Maturitas* 128 (2019): 64–69; Y. Uchida, S. Sugiura, Y. Nishita, N. Saji, M. Sone, and H. Ueda, "Age-Related Hearing Loss and Cognitive Decline: The Potential Mechanisms Linking the Two," *Auris Nasus Larynx* 46, no. 1 (February 2019): 1–9, https://doi.org/10.1016/j.anl.2018.08.010.

11. Stutter and stammer are synonyms, with "stutter" being used more in the United States and "stammer" in the United Kingdom. I prefer "stutter," since it may imply a stopped utterance.

12. Claypole, *Words Fail Us*.

13. Claypole, *Words Fail Us*, 6–7.

14. Claypole, *Words Fail Us*, 14.

15. Claypole, *Words Fail Us*, 206.

16. Claypole, *Words Fail Us*, 88.

17. Quoted from Claypole, *Words Fail Us*, 212.

18. Reminiscent, incidentally, of abductor spasmodic dysphonia.

19. Claypole, *Words Fail Us*, 242.

20. Claypole, *Words Fail Us*, 243–245.

21. Palin has been very supportive of stuttering and helped set up a UK NHS center to support those who stutter. See https://michaelpalincentreforstammering.org.

22. Claypole, *Words Fail Us*, 159.

23. Quoted in Claypole, *Words Fail Us*, 76, from P. Campbell, C. Constantino, and S. Sampson, eds., *Stammering Pride and Prejudice: Difference Not Defect* (Havant: J & R Press, 2019).

24. Claypole, *Words Fail Us*, 90. Quoted from E. Goffman, *The Presentation of Self in Everyday Life* (New York: Doubleday, 1956).

25. Claypole, *Words Fail Us*, 299.

26. Of course, thoughts may be expressed in nonlinguistic domains, in music, or in choreography, but their content in terms of meaning or semantic content is then less than their affective component. Language alone may allow that. (Mathematics enables shared information, but not usually in relation to emotional or affective components.)

27. It is not always the case after aphasia, either. Remember how Faye recalled her odd "Steffi Graf" foreign language syndrome, and still finds a difference between her real voice and the internal one, based as it is on her pre-stroke voice.

28. D. M. Wolpert, K. Doya, and M. Kawato, "A Unifying Computational Framework for Motor Control and Social Interaction," *Philosophical Transactions of the Royal Society B* 358, no. 1431 (March 2003): 593–602.

29. M. J. Pickering and S. Garrod, "An Integrated Theory of Language Production and Comprehension," *Behavioral Brain Sciences* 36, no. 4 (August 2013): 329–347.

30. See postscript for a discussion of empirical work in this area.

31. One woman with Möbius described how she also developed emotional experi-ence and expression as a young adult, in her case while she was studying at a music school. Rehearsing opera she learned to feel the emotion within the song. Occasion-ally, however, she used to find herself feeling the wrong emotion for a given situa-tion, wanting to laugh inappropriately.

32. Claypole, *Words Fail Us*, 67–70.

33. Claypole, *Words Fail Us*, 69–71.

34. M. Greenfield and E. S. Ganzfried, *The Word Escapes Me: Voices of Aphasia* (Bloomington, IN: Balboa Press, 2016), 21.

35. This area is close to post-traumatic growth theory; see R. G. Tedeschi and L. G. Calhoun, "Posttraumatic Growth: Conceptual Foundations and Empirical Evi-dence," *Psychological Inquiry* 15, no. 1 (2004): 1–18, https://doi.org/10.1207/s15327 965pli1501_01.

36. There is a huge literature on conversational analysis; for example, see J. Sid-nell and T. Stivers, eds., *The Handbook of Conversation Analysis* (Chichester: Wiley-Blackwell, 2013), 59–76. For an early paper, see H. Sacks, E. A. Schegloff, and G. Jefferson, "A Simplest Systematics for the Organization of Turn-Taking for Conversa-tion," *Language* 50, no. 4 (1974): 696–735.

37. A. P. Leff, S. Nightingale, B. Gooding, J. Rutter, N. Craven, M. Peart, A. Dunstan, et al., "Clinical Effectiveness of the Queen Square Intensive Comprehensive Aphasia Service for Patients with Poststroke Aphasia," *Stroke* 52 (2021): e594–e598, https://doi.org/10.1161/STROKEAHA.120.033837. Their paper reviews evidence from meta-analyses on the effectiveness of speech therapy, but this suggests that around 100 hours per person is required, more than the health system in the United Kingdom can usually provide. Leff et al. have therefore developed an intensive course of speech therapy, for seven hours per day, five days per week, for three weeks, with a team of speech and language therapists, speech and language therapist assistants, neuropsychologists, and a neurologist. In this way they can reduce the cost to one commensurate with some other interventions and so feel it is a model of service delivery that could be adopted more widely.

38. M. C. Brady, M. Ali, K. VandenBerg, L. J. Williams, L. R. Williams, M. Abo, F. Becker, et al., "Complex Speech-Language Therapy Interventions for Stroke-Related Aphasia: The RELEASE Study Incorporating a Systematic Review and Individual Par-ticipant Data Network Meta-Analysis," *Health and Social Care Delivery Research* 10, no. 28 (2022). For speech and language therapy for aphasia after a stroke, see "Early, Intense Therapy for Language Problems after a Stoke Is Linked to the Greatest Bene-fits," National Institute for Health and Care Research (NIHR), https://evidence.nihr .ac.uk/alert/therapy-for-language-problems-after-a-stroke-is-most-effective-when

-given-early-and-intensively. For a recent comment, see H. Saul, S. Cassidy, B. Deeney, C. Imison, and M. Brady, "Early, Intense Therapy for Language Problems after a Stroke Is Linked to the Greatest Benefits," *BMJ* 383 (2023): 2560, https://doi .org/10.1136/bmj.p2560.

39. "Disfigurement" was his preferred term; J. Partridge, *Face It: Facial Disfigurement and My Fight for Face Equality* (London: Pebble Press, 2020).

40. K. Murphy, *You're Not Listening: What You're Missing and Why It Matters* (London: Harvill Secker, 2020), 32.

41. Murphy, *You're Not Listening*, 38.

42. A. Katz, book review of J. Cole, *About Face*, in *Applied Cognitive Psychology* (1998).

43. Murphy, *You're Not Listening*, 40.

44. Murphy, *You're Not Listening*, 152.

45. J. Cole, *Chekhov's Sakhalin Journey: Doctor Humanitarian Writer* (London: Bloomsbury, 2023).

46. G. Saunders, *A Swim in the Pond in the Rain (in Which Four Russians Give a Masterclass on Writing, Reading and Life)* (London: Bloomsbury, 2012), 335.

47. I. McEwan, quoted in *The Guardian*, September 15, 2001.

48. A. Kafer, *Feminist Queer Crip* (Bloomington: Indiana University Press, 2013).

49. Angela Giuffrida, "'If You Talk, You Live Well': The Remote Sardinian Village with Eight Centenarians," *The Guardian*, August 8, 2021, https://www.theguardian .com/world/2021/aug/08/if-you-talk-you-live-well-the-remote-sardinian-village -famed-for-longevity.

Postscript

1. L. Vygotsky, *Thought and Language*, ed. and trans. Alex Kozulin. (Cambridge, MA: MIT Press, 1986). Original Russian edition published in 1934.

2. C. Kronk, "Private Speech in Adolescents," *Adolescence* 29 (1994): 781–804; C. Fernyhough and F. Fradley, "Private Speech on an Executive Task: Relations with Task Difficulty and Task Performance," *Cognitive Development* 20 (2005): 103–120, https://doi.org/ 10.10016/j.cogdev.2004.11.002.

3. Vygotsky, *Thought and Language*, 210ff.

4. Vygotsky, *Thought and Language*, 210ff.

5. Vygotsky, *Thought and Language*, 210ff.

6. B. Alderson-Day and C. Fernyhough, "Inner Speech: Development, Cognitive Functions, Phenomenology and Neurobiology," *Psychological Bulletin* 141, no. 5 (2015): 931–965, http://dx.dot.org/10.1037/bul0000021.

7. B. Alderson-Day and C. Fernyhough, "Inner Speech."

8. O. W. Sacks, *Seeing Voices* (London: Picador, 1989), 72.

9. R. Passingham, *What Is Special about the Human Brain?* (Oxford: Oxford University Press, 1997), 8–31.

10. M. Muller, quoted from T. Cox, *Now You're Talking* (London: Bodley Head, 2018), 11.

11. Quote taken from S. Hales, *The Man Who Lost His Language* (London: Jessica Kingsley, 2007), 199.

12. S. Pinker, *The Language Instinct* (New York: William Morrow, 1994), 70.

13. O. W. Sacks, *Musicophilia* (London: Picador, 2007); D. Levitin, *This Is Your Brain on Music: Understanding a Human Obsession* (London: Penguin, 2019).

14. One of the joys of Siobhan Davies's work is to see how her pieces reveal thought through action and dance in a way completely independent of cognitive linguistic expression: beautiful, embodied, kinetic melodies.

15. Pinker, *The Language Instinct,* 78ff.

16. A colleague, Chris Miall, once cowrote a paper, "Auguste Rodin Draws Blind." This could mean that Rodin was blind but still drew, or that he drew people who were blind, or that he made a drawing of his window blinds, or that simply he drew those blinds shut. In fact, it meant that Rodin, toward the end of his career, used to draw sketches of dancers without seeing what he was drawing. Five distinct meanings from three words, (the given name being redundant to the point). See J. Tchalenko and R. C. Miall, "Auguste Rodin Draws Blind: An Art and Psychological Study," *Leonardo* 52, no. 5 (2019): 483–491.

17. Pinker, *The Language Instinct,* 81.

18. While Pinker takes several pages to demolish the idea, it only ever received a footnote in Sacks's *Seeing Voices.*

19. B. Alderson-Day and C. Fernyhough, "Inner Speech."

20. I was once at Waterloo Station in London late at night waiting for a train. On the concourse was a group of Deaf teenagers signing. A young girl started to argue with her boyfriend. It became operatic, as sign and gesture soared ever upward in size and expression. It ended with the girl storming off, fully fifty meters away, before she stopped, with back turned, arms folded in disgust. Then—inevitably—she looked back over her shoulder to check what was happening. The performance

was magnificent and far more engaging (observed carefully, of course) than a mere slanging match in sound.

21. For discussion, see H. Knoors and M. Marschark, "Language Planning for the 21st Century: Revisiting Bilingual Language Policy for Deaf Children," *Journal of Deaf Studies and Deaf Education* 17, no. 3 (November 2012): 291–305, https://doi .org/10.1093/deafed/ens018.

22. See K. Zimmermann and P. Brugger, "Signed Soliloquy: Visible Private Speech," *Journal of Deaf Studies and Deaf Education* 18, no. 2 (April 2013): 261–270, especially 263, https://doi.org/10.1093/deafed/ens072.

23. Zimmerman and Brugger, "Signed Soliloquy."

24. P. Delamillieure, G. Doucet, B. Mazoyer, M.-R. Turbelin, N. Delcroix, E. Mellet, L. Zago, et al., "The Resting State Questionnaire: An Introspective Questionnaire for Evaluation of Inner Experience during the Conscious Resting State," *Brain Research Bulletin* 81, no. 6 (April 2010): 565–573, https://doi.org/10.1016/j.brainresbull.2009 .11.014.

25. S. Geva. P. S. Jones, J. T. Crinion, C. J. Price, J.-C. Baron, and E. A. Warburton, "The Neural Correlates of Inner Speech Defined by Voxel-Based Lesion-Symptom Mapping," *Brain* 134 (October 2011): 3071–3082, https://doi.org/10.1093/brain /awr232; S. Geva, S. Bennett, E. Warburton, and K. Patterson, "Discrepancy between Inner and Overt Speech: Implications for Post-Stroke Aphasia and Normal Language Processing," *Aphasiology* 25 (2011): 323–343, https://doi.org/10.1080/02687038 .2010.511236.

26. L. Vercueil and M. Perrone-Bertolotti, "Ictal Inner Speech Jargon," *Epilepsy & Behavior* 27 (2013): 307–309, https://doi.org/10.1016/j.yebeh.2013.02.007.

27. Vercueil and Perronne-Bertolotti, "Ictal Inner Speech Jargon," 308 (emphasis added).

28. P. Langland-Hassan, F. R. Faries, M. J. Richardson, and A. Dietz, "Inner Speech Deficits in People with Aphasia," *Frontiers in Psychology* 6 (2015), https://www.fron tiersin.org/articles/10.3389/fpsyg.2015.00528.

Acknowledgments

1. J. Cole, *Pride and a Daily Marathon* (London: Duckworth, 1991; Cambridge, MA: MIT Press, 1995); J. Cole, *About Face* (Cambridge, MA: MIT Press, 1998); J. Cole, *Still Lives* (Cambridge, MA: MIT Press, 2004); J. Cole with H. Spalding, *The Invisible Smile* (Oxford: Oxford University Press, 2008); J. Cole, *Losing Touch* (London: Oxford University Press, 2016).

2. Cole, *About Face.*

Index